Contractor's Project Guide
To Public Agency Contracts

EDWARD R. FISK, PE
Consulting Construction Engineer
Licensed General Contractor

JAMES R. NEGELE, J.D., BSCE
Attorney at Law
Negele, Knopfler, Pierson, & Robertson

JOHN WILEY & SONS
New York   Chichester   Brisbane   Toronto   Singapore

DISCLAIMER

The opinions and recommendations expressed in this book are those of the authors and reflect their knowledge and expertise gained through years of practical experience. Any reference to laws or legal theories are, by necessity, generalizations of the status of the law throughout the United States as a whole. To address each and every state's and jurisdiction's statutes and case law relating to public works contracting is beyond the scope of this book. Consequently, although construction law and specific laws affecting construction generally have been addressed in this book, it should in no way be used as a substitute for advice from sound legal counsel. When a problem with legal overtones arises, it is strongly recommended that the contractor consult with competent legal counsel as to the status of the law in the particular jurisdiction involved.

Figures 6.1, 6.3, 9.2–9.6, 9.8, 9.9, 9.11, 9.12, and 9.14 are forms reprinted from *Construction Engineer's Form Book,* Edward Fisk. Copyright © 1981, John Wiley & Sons. Reprinted by permission.

Figures 4.1–4.7 and 10.1 are from *Construction Project Administration,* 2nd ed., Edward Fisk. Copyright © 1983, John Wiley & Sons. Reprinted by permission.

Copyright © 1988, by John Wiley & Sons, Inc.

All rights reserved. Published simultaneously in Canada.

Reproduction or translation of any part of
this work beyond that permitted by Sections
107 and 108 of the 1976 United States Copyright
Act without the permission of the copyright
owner is unlawful. Requests for permission
or further information should be addressed to
the Permissions Department, John Wiley & Sons.

Library of Congress Cataloging in Publication Data:
Fisk, Edward R.
   Contractor's project guide to public agency
contracts.

   Bibliography: p. 333
   Includes index.
   1. Building—Contracts and specifications—United
States.   2. Public contracts—United States.
3. Construction industry—Law and legislation—
United States.   I. Negele, James R.   II. Title.
KF865.F57   1988   346.73′023   87-10455
ISBN 0-471-88873-7   347.30623

Printed in the United States of America

10 9 8 7 6 5 4 3 2 1

PREFACE

To a new contractor or an established contractor that previously has worked only in the private sector, the prospect of doing work for a public agency may seem like an ideal opportunity to expand into a profitable construction environment. Governmental agencies always need work done, and there is usually no question as to their financial stability. Although construction per se is much the same in both the private and public sectors, the bidding before a job is awarded and the administration of a construction contract vary dramatically from the public sector to the private sector.

As nearly every aspect of public works contracting is dictated by various laws and regulations, the contract administration responsibilities of the contractor are substantially greater than in the private sector. Unless a contractor is aware of the way that public works contracts function and of their administrative requirements, a public works contract can spell disaster for the uninformed contractor.

PREFACE

This book is about construction and is addressed to general and prime contractors involved with all aspects of construction on public works projects. Here, the word "contractor" has been used to refer to a general contractor or a prime contractor only, but not a subcontractor. The word "project" has been used to refer to any type of public works construction project.

Although many principles discussed apply to construction in the private sector as well, all of our discussions are directed to construction work being done for governmental agencies at all levels, for the benefit of the public, and with public funds. The word "agency" refers to a public agency, government agency, contracting agency, federal agency, state agency, local agency, and owner, as the principles discussed apply to all of them.

It is our hope that this book can help contractors planning to work in the public sector by alerting them to some of the problems that can be expected when working with a public agency.

Mr. Fisk is a licensed general contractor and a registered professional engineer and land surveyor in 12 states with over 38 years of experience in both private and public works construction. Mr. Negele is a practicing trial lawyer specializing in construction disputes, with degrees in both civil engineering and law. We have worked together professionally on construction claims and now bring our combined experience together in book form to enable contractors new to the public sector to avoid the pitfalls that we have so often seen befall the unwary contractor.

Grateful acknowledgment is made of the valuable contributions of those who carefully reviewed, critiqued, and offered suggestions for improving this book. In particular, we wish to thank James Acret, Esq., Attorney at Law, Santa Monica, California; Neal B. H. Benjamin, P.E. and Professor of Civil Engineering, University of Missouri–Columbia, Missouri; John D. Borcherding, Associate Professor of Civil Engineering, University of Texas, Austin, Texas; Joseph Litvin, Esq., P.E., Attorney at Law, Dayton, Ohio; Bernard P. Monahan, P.E. and Contractor, Queens County, New York, N.Y.; Robert A. Rubin, Esq., P.E., Attorney at Law, New York, N.Y.; and Clifford Schexynayder, P.E. and Adjunct Professor, Department of Civil Engineering, Louisiana Technical University, Ruston, Louisiana.

EDWARD R. FISK, P.E.
Orange, California
JAMES R. NEGELE, ESQ.
Los Angeles/Universal City, California

August 1987

CONTENTS

CHAPTER 1 — **CONTRACTING FOR PUBLIC WORKS PROJECTS** — 1

TYPES OF CONTRACTS — 5
- Lump Sum Contract 5
- Unit Price Contract 6
- Cost-Plus Contract 7
- Incentive Contract 7

PARTNERSHIPS AND JOINT VENTURES — 8

CONSTRUCTION CONTRACTS WITH PUBLIC AGENCIES — 9

CHAPTER 2 — **PUBLIC WORKS vs. PRIVATE CONTRACTS** — 11

WHY A DIFFERENCE? — 12

MAKING IT WORK FOR YOU — 13

	PRINCIPAL DIFFERENCES	14
	Product Substitutions 15	
	Protection for Subcontractors 17	
	Public Advertising of a Project 17	
	Bonding Requirements 17	
	Relief from Bid Errors 19	
	Documentation 19	
	CONTRACTUAL STATUS OF CONTRACTORS AND SUPPLIERS	20

CHAPTER 3 RISK ALLOCATION — 21

RISK AND LIABILITY SHARING	24
DEFINITION OF RISK	24
SCOPE AND APPLICABILITY TO CONSTRUCTION	25
IDENTIFICATION AND NATURE OF CONSTRUCTION RISKS	26
SHIFTING OF RISK BY CONTRACT	26
Exculpatory Clauses 27	
WHO SHOULD ACCEPT WHAT RISKS?	29
MINIMIZING RISKS AND MITIGATING LOSSES	29
TYPES OF RISKS AND ALLOCATION OF THOSE RISKS	29
CONTRACTUAL ALLOCATION OF RISK	37
Reallocation of Risks to Subcontractors 39	
RISKS RESERVED TO THE CONTRACTOR	40
CONTRACTOR PARTICIPATION IN VALUE ENGINEERING	41
DISPUTES	42

CHAPTER 4 SPECIFICATIONS AND DRAWINGS — 43

WHAT IS A SPECIFICATION?	44
CONFLICTS BETWEEN DRAWINGS AND SPECIFICATIONS	44
CONTENT AND COMPONENT PARTS OF A SPECIFICATION	45

CONTENTS

Content of the Specifications 45
Component Parts of a Specification 47

WHAT DO THE SPECIFICATIONS MEAN TO THE CONTRACTOR? 51

CSI SPECIFICATIONS FORMAT: ITS MEANING AND IMPORTANCE 52
CSI 16-Division Format 52
CSI Division/Section Concept 56
CSI Three-Part Section Format 56

STATE HIGHWAY DEPARTMENT FORMATS 57
AASHTO Standard Format for Highway Construction Specifications 58
California Department of Transportation Format 60

NONSTANDARD CONSTRUCTION SPECIFICATION FORMATS IN USE 61

PROJECT SPECIFICATIONS vs. SPECIAL PROVISIONS CONCEPT 62
Project Specifications (CSI Project Manual) 62
Special Provisions or Supplemental Specifications 63

GENERAL CONDITIONS OF THE CONSTRUCTION CONTRACT 64
General Conditions Provisions of Standard Specifications 66

SUPPLEMENTARY GENERAL CONDITIONS 67

TECHNICAL PROVISIONS OF THE SPECIFICATIONS 68
Provisions for Temporary Facilities 70

ADDENDA TO THE SPECIFICATIONS 71

STANDARD SPECIFICATIONS 73

SPECIAL MATERIAL AND PRODUCT STANDARDS 75
Government Standards 76
NonGovernment Standards 79

ACCESS TO SPECIAL STANDARDS BY CONTRACTOR 83

BUILDING CODES, REGULATIONS, ORDINANCES, AND PERMITS 84
Projects Subject to Control by More Than One Agency 86

TYPES OF DRAWINGS COMPRISING THE CONSTRUCTION CONTRACT 86
Contract Drawings 86
Standard Drawings 86
Shop Drawings 87
Record Drawings/As-Built Drawings 88

CHAPTER 5   CONSTRUCTION LAWS   89

- COMPLIANCE WITH LAWS AND REGULATIONS — 91
- PURPOSE OF CONSTRUCTION LAWS — 92
- LIMITATIONS OF AUTHORITY OF A PUBLIC AGENCY — 93
- PRINCIPAL TYPES OF LAWS GOVERNING PUBLIC WORKS CONSTRUCTION — 94
- PUBLIC vs. PRIVATE CONTRACTS — 96
- TRAFFIC REQUIREMENTS DURING CONSTRUCTION — 97
- WORK WITHIN OR ADJACENT TO NAVIGABLE WATERWAYS — 99
- SUBCONTRACTS — 101
  - Bid Shopping or Bid Peddling   102
  - Effect of Bid Shopping   103
  - Control of Bid Shopping   104
  - Protective Legislation for Subcontractors   104
- LABOR LAWS — 105
  - The Sherman Anti-Trust Act   106
  - The Landrum–Griffin Act   106
  - The Equal Opportunity Employment Act   106
  - Executive Order 11246   107
  - The Philadelphia Plan   107
  - The Davis–Bacon Act   107
  - The Copeland Act   108
  - The Hobbs Act   109
  - The National Apprenticeship Act   109
- ETHNIC MINORITIES AND WOMEN IN CONSTRUCTION — 109
- LABOR RELATIONS — 110
- COLLECTIVE BARGAINING — 110
- PREJOB LABOR CONSIDERATIONS — 111
- OPEN-SHOP CONTRACTING — 112
- WORKER'S COMPENSATION AND EMPLOYER LIABILITY INSURANCE — 113
- LICENSING LAWS — 114

CHAPTER 6 — CONSTRUCTION SAFETY — 115
- BASIC RESPONSIBILITY FOR ON-SITE SAFETY — 116
- PRIMARY LIABILITY — 116
  - Subcontractor Involvement with Safety 117
  - Subcontractors Should Be Monitored 117
- OSHA AND THE CONTRACTOR — 117
  - Safety Plans 119
  - Record Keeping 121
  - Safety Violation Procedures 121
- CONTRACTUAL SAFETY REQUIREMENTS — 124
- ACCIDENT RECORDS — 125
- PROTECTION OF THE PUBLIC — 128
- TRENCH AND EXCAVATION SAFETY — 129

CHAPTER 7 — BIDDING PHASE — 133
- SELECTION OF SUBCONTRACTORS — 135
- PREBID INSPECTIONS AND CONFERENCES — 136
- DISCLAIMERS IN BID SOLICITATION — 137
- PREPARATION AND SUBMITTAL OF BIDS — 138
  - Qualifying of Bids 138
  - Responsive Bids 139
  - Unit Price vs. Lump Sum Bids 140
  - Time of Submittal of Bids 142
  - Bid Bonds 143
  - Bid Errors—Relief of Bidders 144
  - Opening of Bids 145
- EFFECTS OF WEATHER ON BID PRICE — 145
- NEGOTIATION OF BID PRICES — 146
- PRODUCT SUBSTITUTIONS — 147

CONTENTS

OWNER DISCLOSURE OF SITE INFORMATION	148
SCHEDULE OF VALUES	149

CHAPTER 8 PRECONSTRUCTION PHASE — 151

AWARD AND CONTRACT	152
INSURANCE; BONDS; PERMITS	152
SUBCONTRACTING	154
Time for Performance 154	
Incorporation of Prime Contract 155	
Key Provisions for Subcontracts 156	
PRECONSTRUCTION CONFERENCE	158
Definitions 159	
Purpose 160	
Time for the Conference 160	
Subjects for Discussion 160	
Agenda Items for a Typical Preconstruction Conference 161	
NOTICE TO PROCEED	163
SCHEDULE OF VALUES ON LUMP-SUM JOBS	164

CHAPTER 9 PROJECT ADMINISTRATION — 167

THE CONTRACT	168
Contract Documents 168	
Prescriptive vs. Performance Specifications 168	
ADMINISTRATION	170
Supervision of the Work 170	
One-to-One Concept 170	
The Contractor and Subcontractors 171	
Submittals of Shop Drawings and Samples 172	

DOCUMENTATION	177
Records and Record Keeping 178	
Records Required by the Contract 178	
Records for Claims and Disputes 179	
Types of Records for Claims Support 180	
Employee Time Records 180	
Job Costs 180	
Progress Reporting 180	
On-Site Relations or Transactions 182	
Time and Cost Monitoring 185	
Requests for Services 189	
TEMPORARY FACILITIES	189
Field Offices 190	
Communications 190	
Environmental Controls 190	
Project Signs 191	
SCHEDULING AND COORDINATION	192
Schedule Submittals 193	
Form of the Schedule 193	
CPM and Cost Reporting 194	
CONTRACT TIME	194
"Time is of the Essence" 195	
Stopping or Suspension of the Work by the Architect-Engineer 195	
DIFFERING SITE CONDITIONS	195
Unforeseen Underground Conditions 196	
CHANGE-ORDER ADMINISTRATION	197
Factors the Agency Will Take into Consideration 201	
Change Orders and Extra Work in Subcontracts 201	
Field Order or Work Directive Change 204	
JOB CHANGES AND PERFORMANCE BONDS	204
LABOR AND MATERIAL RELEASES	205
PAYMENT	206
MATERIALS AND METHODS	207
Handling Rejected Work 207	

Nonconforming Work 207
Substitution of Products 208

CHAPTER 10 VALUE ENGINEERING DURING CONSTRUCTION 209

DEFINITION OF VALUE ENGINEERING 211

FUNDAMENTALS OF VALUE ENGINEERING (VE) 213
Function 211
Worth 212
Cost 212
Value 212
The Philosophy of Value 213
Types of Value Engineering Recommendations 213

VALUE ENGINEERING BY THE CONTRACTOR 213

METHODOLOGY IN GENERATING VE PROPOSALS 215
Information Phase 217
Speculation Phase 218
Analysis Phase 218
Development Phase 219
Presentation and Followup Phase 220

CHAPTER 11 CONSTRUCTION CLAIMS 223

CLAIMS AND DISPUTES 225
Contractor Must Alert Agency 225
Administrative Procedures 225
Work Performed Under Protest 226
Agency Must Have Opportunity to Correct 226
Preparation of a Claim File 227
The Right to File Claims 227

LEGAL COUNSEL 228
Involvement from the Project Beginning 229

DIFFERENCES BETWEEN THE PARTIES 229
Owner-Caused Delays 229
Home Office Overhead 232

Flaws in Use of the Various Computation Methods 238
Scheduling Changes 239
Constructive Changes 240
Differing Site Conditions 241
Unusually Severe Weather Conditions 242
Acceleration of the Work by the Public Agency/Owner 243
Suspension of the Work; Termination 243
Failure to Agree on Change Order Pricing 245
Conflicts in Plans and Specifications 246
Miscellaneous Problems 247

SUBCONTRACTOR RELATIONS   247
Dispute Clauses 247
Damages for Delay 249
Unforeseen or Unusual Site Conditions 250
Claims on Subcontractors' Bonds 250

DOCUMENTATION FOR CLAIMS   250

RESOLUTION OF CONSTRUCTION DISPUTES   253

SUPPORTING THE CONTRACTOR'S POSITION   254
Burden of Proof 255
Method of Presentation 255
Evidence 255

ALTERNATIVE METHODS FOR DISPUTE SETTLEMENT   258

CHAPTER 12 PROJECT CLOSEOUT   259

COMPLETION OF THE WORK   260

PUNCH LIST PROCEDURES   260
Work by the Contractor and Subcontractors 260
Work by the Contractor 261
Time of Prefinal Inspection 261
Preparation of the Punch List 261
Final Punch List Inspection 264

CLEANUP OF THE PROJECT SITE   264

CONTENTS

FINAL SUBMITTALS	265
BENEFICIAL USE OR PARTIAL UTILIZATION	265
SUBSTANTIAL COMPLETION	267
FINAL PAYMENT	269
Withheld Funds from Final Payment 269	
Final Payment and Waivers of Liability 270	
RELEASE OF RETAINAGE	271
LIQUIDATED DAMAGES FOR DELAY	271
Cause of delay 272	
Contractor's Rights 273	
Liquidated Damages Are Not a Penalty 273	
Liquidated Damages Clauses in Subcontracts 274	
Amount of Liquidated Damages 274	
"Time Is of the Essence" 274	
LIENS AND STOP-PAYMENT ORDERS	275
Contractor Protection 275	
Lien Waivers 276	
Subcontractor Lien Releases 276	
PICKUP WORK	277
Uncompleted Punch List Items 277	

APPENDICES   279

Appendix A	Surety and Insurance Bond Checklist	279
Appendix B	Construction Industry Arbitration Rules—American Arbitration Association	291
Appendix C	Sample Public Works Contract Documents (Contract Documents for Construction of Federally Assisted Water and Sewer Projects)	303

BIBLIOGRAPHY   333
INDEX   336

CHAPTER 1

Contracting for Public Works Projects

- Types of Contracts
- Partnerships and Joint Ventures
- Construction Contracts with Public Agencies

CHAPTER 1 CONTRACTING FOR PUBLIC WORKS PROJECTS

This is not a book on how to manage your construction projects. That task is up to you. This is a guidebook to help you sidestep the usual pitfalls when expanding your business from private contracts to public works contracting. The authors make no pretense of offering management methods that will revolutionize your business. Only, that if you follow our guidelines, you will not be caught unaware when you suddenly realize that you failed to include enough money in your bid to cover the many unforeseen costs commonplace in public works construction contracts.

The newcomer to the game of public works contracting may be in for a shocking surprise if the work is judged by private work of a similar nature. Those appealing advertisements in the daily construction newspapers and the local newspapers seem to suggest that there is a virtual pot of gold at the end of the construction rainbow, and that all you need do to reach it is to underbid the competition, and the rest should be all downhill. If anything is true, it is the opposite. That is precisely where the contractor's trouble begins. The methods used in managing the construction projects that worked so well in the private sector cannot be readily applied to the much more regimented and documented area of public works contracting. The old concept of simply providing the agency with a "good job" is no longer sufficient. The new game must be played "by the book."

It all begins with the Notice Inviting Bids or another similarly named document. The bidders should be prepared to follow all instructions to the letter. The often followed practice among contractors in the private sector of submitting a "qualified" bid by marking over portions of the bidding documents serves little in public works contracting, except to have the entire bid rejected as being "nonresponsive" without any opportunity for the bidder to negotiate or make corrections. Then all of the time and money spent by the contractor to prepare the bid is needlessly wasted. There are no second chances. The bid must be right the first time, as there will be no opportunity to talk about it later on.

In the game of public works contracting, the bidder had better be prepared to read *all* of the specifications, including the front-end documents *before* submitting its bid; otherwise, later surprises may be sufficient to force a small contractor into bankruptcy. Some contractors refer to public bidding as "a method of finding a contractor willing to work at a loss." Interpretations of apparent ambiguities in the contract documents can be the source of serious problems. The traditional legal concept that in the case of ambiguity, a contract is to be interpreted in favor of the party who did not draft the documents is often viewed in reverse when dealing with a public agency. In public contracts, the documents are considered contracts of "adhesion"; this simply means that the agency writes the document and all of its terms—in its own favor, of course—and your only choice is to take it or leave it, without modification or a chance to negotiate a compromise.

If the new entrant into the field of public works contracting assumes that the contractor's mission is just to build a project as it appears on the plans and specifications, then it is time to review the facts. The project that is supposed to be built may, after awhile, seem to be treated as a secondary byproduct by the agency, and the contractor may soon feel that the true objective is to generate reams of paperwork and make work for local politicians. The paperwork is, indeed, overwhelming at times, but it is important. In fact, as will be shown later in this book, it may even be well for the contractor to initiate a few inhouse paperwork systems of its own. Otherwise, when you want to file claims for recovery of job costs, your case may be lost even before it is ready to present.

Many newcomers to the contracting business feel that they are fairly shrewd people and may feel that they know a few tricks of their own. In all too many cases, the opportunity to use some of these innovative methods may never arrive, as the die may have been cast the moment those voluminous contract documents were signed, without your taking the time to study the terms as carefully as you should have. Generally, those contract documents were compiled by experts whose only objective (it sometimes seems) was to place all of the risk on the contractor's shoulders and none on the public agency.

That is what this book is all about. It will help you cope with the public agency and the architect-engineer in their own game. You will be guided through many of the high risk areas that lie in wait for the unsuspecting contractor who, on a first public works contract, naively bids far too low, not truly realizing what to expect. You will not only be shown how to avoid many of the difficulties that await you, but your legal position on many types of claims will be discussed.

It should be remembered that the architect-engineer and the agency are not generally out to "get" the contractor, but are just trying to protect their own positions (though sometimes excessively). They are usually just following the numerous laws governing the administration of public works contracts at all levels. A public agency may not be aware of a newcomer's unfamiliarity with the special circumstances of public works contracting and merely treats each contractor in the same manner as it would any other of the more experienced public works contractors. The agency is required by law to administer its contracts in a fashion that will provide maximum protection to the public, not the contractor.

CHAPTER 1 CONTRACTING FOR PUBLIC WORKS PROJECTS

The solution is to not try to fight city hall by attempting to bypass the rules. The approach will only cost both ill-will and your money. Learn to fight back by playing the game by their own rules and winning. In this book, the authors suggest many ways of doing just that. Although we do not cover *all* the angles, enough is presented to keep the average contractor solvent and out of trouble if it follows the advice offered.

TYPES OF CONTRACTS

There are a variety of types of contracts used in the construction industry. They include the following.

Lump Sum Contract

Among the most commonly used types of contracts by public agencies, lump sum contracts are particularly suited for projects that can be reasonably well defined, evaluated, and bid. The construction or remodeling of a police facility, wastewater treatment plant (see Figure 1.1), hospital, and most buildings are projects of the type that would normally require a lump sum contract.

FIGURE 1.1 City of Escondido-Vista Irrigation District, 75 MGD (ultimate) water treatment plant.

The contract price in a lump sum bid is the bottom line total, even when a public agency requires the submittal of a bid breakdown or "schedule of values."

Unit Price Contract

A unit price contract is very commonly used by public agencies for civil engineering construction such as roads or street improvements, pipelines (see Figure 1.2), underground utility lines, canals, dams, and projects involving earthworks or hydraulic structures.

In such contracts, the unit prices quoted for the various items of work,

FIGURE 1.2 Las Vegas Water District, Sandhill–Lamb pipeline. (*Source*: Photo courtesy of J.M. Montgomery Consulting Engineers, Inc.)

such as cubic yards of excavation, backfill, or embankment; linear feet of pipeline; square feet, tons, or cubic yards of pavement materials laid; linear feet of chain link fencing; and similar items, serve as the basis for determination of the contract price. The *price used for comparing bids* for the award of a contract will be the unit price bid by the contractor multiplied by the quantities listed in the engineer's estimate; the *final prices paid* to the contractor will be determined by the actual quantities of materials actually placed or indicated on the drawings multiplied by the quoted unit price. The "contract price" in a unit price bid is not the bottom line total of the estimate extensions, but is actually the unit prices bid.

Cost-Plus Contract

A cost-plus contract may be either a cost-plus-percentage-of-cost contract, which is actually very similar to a time and materials contract, or a cost-plus-fixed-fee contract. With the cost-plus-percentage-of-cost type contract, the contractor will be reimbursed for its direct costs and will receive, in addition, a percentage of that cost. With a cost-plus-fixed-fee contract, which is more frequently used with design professionals, but occasionally with contractors as well, the contractor would receive a base amount to perform the work. The base amount would include all overhead and profit, and the contractor would then be reimbursed for all of its costs.

Incentive Contract

Although incentive contracts have not been widely used in public works contracting, they can be very effective under the right circumstances. An incentive contract provides the contractor with an opportunity for greater financial gain based upon performance and ability to meet certain preestablished goals. With the increased and more widespread use of construction contracts containing liquidated damages clauses for delay, incentive contracts for such factors as early completion can provide the public agency with the possibility of saving money, and the contractor with the possibility of increased profits by performing in a certain manner.

Of all the types of contracts listed, the two most commonly encountered and to which the remainder of the book will generally be directed are lump sum contracts and unit price contracts.

PARTNERSHIPS AND JOINT VENTURES

A partnership is an association of two or more entities conducting business for profit. The principal benefit to contractors in partnership is the concentration of assets, equipment, facilities, and human resources. By joining together, such entities can pool their financial resources and obtain an increased bonding capacity that will allow the partnership greater job opportunities. Generally, each partner makes a contribution to the capital or assets in creating the partnership. Profits and losses are usually shared in the same proportion as the percentages of ownership between the partners. Usually, partnerships are created by an agreement between the partners and any provisions or requirements of applicable state law. Except for statutory requirements, the partnership agreements can contain any provisions that the partners believe are necessary. As the formation of a partnership entails the creation of a new entity, it is recommended that a lawyer familiar with partnership law assist in the formation of a partnership.

As a member of a partnership, each entity is not only responsible for itself, but also for each of its partners. Assuming a general partnership, each member is the agent of the partnership and has the authority to bind the partnership, enter into contracts, and act for its other partners in the business. Each partner assumes unlimited personal liability to third parties for all debts of the partnership. In addition, each partner is individually liable for all of the acts of the other parties in the ordinary conduct of the partnership's business affairs. Each partner is jointly and severally liable for partnership debts.

Since a partnership requires that each partner accept unlimited financial responsibility for the acts of his or her partners as well as his or her own acts, careful consideration must be given to the formation of any partnership.

Similar to a partnership, a joint venture can be considered a special-purpose partnership. Like a partnership, it is formed by two or more entities that pool their resources and unite forces through the medium of a joint venture. Joint ventures often are used by contractors to pool facilities and personnel and to spread construction risks.

A joint venture relates only to a single project, regardless of how long it will take to complete it. By way of the joint venture, the coventurers combine their resources, assets, and skills to complete that one project. As with partnerships, each joint venture participates in the work and will share any profits or losses. In addition, each coventurer is not only legally liable for the performance of the entire contract, but also jointly and severally liable for all joint venture debts, regardless of the degree of participation.

As in a partnership, a joint venture agreement creates a new entity. As stated before, it is recommended that contractors considering a joint venture secure the services of a lawyer for advice on the legal aspects and liabilities of forming a joint venture for each coventurer.

CONSTRUCTION CONTRACTS WITH PUBLIC AGENCIES

The actual construction contracts used for public works construction and the terms contained within them vary widely among the various jurisdictions, levels of government, and types of project facility. However, what they do all have in common is the fact that they must comply with the specific laws, regulations, and ordinances for the given type of project in a particular jurisdiction and locale under a particular governmental agency and are subject to the source of funding for the project. In addition to the types of provisions contained in a construction contract for private work, a construction contract for public works must include all of the necessary language and provisions required by applicable local laws.

The one factor all construction contracts for public works have in common is that the work involves the expenditure of public funds. When public funds are involved, the contracts must contain such required provisions in order to protect or further the public interest. All such required contract provisions are included to further some public policy.

CHAPTER 2

Public Works vs. Private Contracts

- Why a Difference?
- Making It Work for You
- Principal Differences
- Contractual Status of Subcontractors and Suppliers

CHAPTER 2 PUBLIC WORKS VS. PRIVATE CONTRACTS

WHY A DIFFERENCE?

Because of the public's interest in all public works projects, whether built with public funds or private funds for the intended use of the public, construction contracts for public works are required to contain provisions that have evolved over time in order to protect and best serve the public interest.

In addition to the necessity of containing specific legal language, public contracts have general requirements that are handled differently than those for private contracts.

For example, public contracts in the United States are generally awarded to the lowest responsive, responsible bidder with very little consideration of the reputation or track record of the lowest bidder. Whereas the lowest bidder can be rejected in a private project if the owner feels that the lowest bidder will not do as good a job as the second or third lowest bidder, a public entity, without real cause, cannot avoid awarding the contract to the lowest responsive, responsible bidder.

In order to ensure that the project in question will be built for the public at the lowest possible cost *and* that all of the interested bidders have an equal opportunity to be awarded the contract, all public works projects (over nominal local values) must be publicly advertised. The Notice Inviting Bids must normally contain a list of prevailing wage rates for certain crafts to be used in the work. When federal funds are involved, a list of federal wage rates must be included. In order to ensure the integrity of the competitive bidding process and protect the public's interest, bids on construction contracts for public works must be accompanied by a Bid Bond, which is usually five or ten percent of the total price cost bid for construction of the project. The bid bond is intended to protect the public in case a low bidder fails to honor its bid by executing a contract to construct the project.

In order to assure the contractor's performance and protect the public interest, the contractor must post a Performance Bond. In addition, to protect the financial interest of the subcontractors and material suppliers involved in the project as well as to protect the project itself, the contractor must post a Payment Bond (for labor and materials) in the amount specified by law. The Performance Bond ensures that the project is completed, while the Payment Bond ensures that the general contractor pays for work performed by subcontractors and for materials used in the project.

In order to protect the public's interest and the government itself, the contractor must obtain insurance policies and bonds covering public liability and property damage.

In order to protect the integrity of the bidding process and the subcontractors who will work on the project, some jurisdictions require that all subcontractors who will perform work on the project be listed and filed with the general contractor's bid. Most public agencies also require that in specifications where a "brand name" product is specified, the specifier must include the names of two such products followed by the words "or equal." The "Or Equal" is intended to ensure the integrity of the competitive bidding system and give all manufacturers the broadest opportunity to supply their products to public works projects.

MAKING IT WORK FOR YOU

The majority of points discussed in this book deal with the administration of contracts with public agencies, primarily because they are the more regulated. In private contracts, the only controls affecting the parties are those to which the parties to the contract have agreed among themselves. Thus, there is room for negotiation of almost every point.

In public works contracting, however, it is felt that the public good is involved—the taxpayer, that is—and someone should look out for the interests of the public body. To accomplish this in such a way that there is uniform application of the principles on all public contracts, various laws and regulations have been enacted and are enforced to protect the public interest. Although many of these laws appear on the surface as bureaucratic red tape, the origins of each of the regulations are founded on good business principles, and if everyone followed them as a matter of business ethics, we would not need that many protective laws or, for that matter, the apparent endless red tape that seems to accompany public contracts. Unfortunately, there are always a few opportunists who lack any kind of ethics and who, with no pangs of conscience, will take any public agency for all they can get, legally or otherwise. It is the few like this who have brought about the current cumbersome administrative procedures as the only means of protecting the taxpayer. Meanwhile, it is the intent of the authors to provide you with guidance for working within the system without sacrificing profitability.

Before bidding on any public project, if a contractor has not had previous experience in this area, it would be wise to consult with those who have experience. The cost of administering a public project is considerably different from that of private jobs. There are many apparent pitfalls, and the surprises that seem to appear at every corner use up more money from the project budget.

PRINCIPAL DIFFERENCES

Not all of the laws concerning public contracts are designed to protect only the public. Some are aimed at protecting contractors as well. A number of them are specifically directed at the protection of subcontractors. Although the protection varies from state to state, some places, such as California, have regulated almost every aspect of public contracts.

The principal differences between public project contracting and private contracts lies in the following subject areas, some of which were mentioned previously.

1. Compliance with "Or-Equal" provisions of a contract; you generally do not have the right for the unilateral substitution of specified products.
2. Selection and identification of subcontractors in the bid—without the right to make a later change.
3. Projects must be publicly advertised and open to all potential bidders.
4. Bonds are required. These include a Bid Bond, Performance Bond, and Payment Bond. In some states, the performance and payment bonds are executed as a single document.
5. Availability of information during the bid phase is restricted.
6. Conformance with the Department of Labor's wage rate determinations for the area in which the work is to be performed.
7. Compliance with requirements for a certain percentage of minorities, women, and disadvantaged business enterprises.
8. "Buy American" requirements limiting the contractor to the use of only American-made products on the job.
9. Bids must be responsive. Crossouts or changes on a bid sheet will usually invalidate it.
10. No negotiation is possible either during the bidding period or after until an agreement is signed; then, it may be used only as a means of addressing change orders or claims.

11 Requirements in the documents that the contractor serve formal notice to the owner on matters relating to claims, protests, or disputes within a specified period of time or be considered as having waived its rights are rigidly upheld by the courts.

12 Retainage is normally prescribed by law and is a necessity that must be considered when budgeting a project's cash flow requirements.

13 Bid mistakes are generally correctable if they are clerical in nature—that is, mistakes in multiplying, addition, and so on, if the basic work sheets used to compute the bid support this claim. Errors of judgment are not excusable, and you may be bound to performance of an unprofitable contract by the owner in such cases.

14 Public contracts are generally much more formal so remember: The documentation requirements and paper shuffling must be addressed and not considered superfluous.

Product Substitutions

The contract documents should be read carefully. The requirements governing the substitution of "or-equal" products varies somewhat from place to place, but in general they are fairly consistent throughout the United States and in many foreign countries. In most cases, a contractor will be allowed a reasonable length of time during which to offer an "or-equal" product for consideration. In a few rare cases, an owner or architect-engineer will attempt to force a contractor to offer any proposed substitutes and all backup information as a part of its bid. This, of course, is patently unfair, as the contractor would be basing its bid price on the lower-cost product submitted and may find that the offered substitute is later determined by the architect-engineer to be unacceptable. Substitutes are more fairly considered after award of the contract.

The usual statements in the documents that indicate the architect-engineers' decision as final seem to be met with different attitudes in different courts. Basically, the courts appear to hold that the architect-engineer is, indeed, the judge of equality; however, the courts' interpretation of "equality" differs widely from that of many owners and their architect-engineers. In the view of the courts, it is generally held that *equal* means *comparable quality and utility* only. That interpretation excludes esthetics, and it precludes any attempt on the architect-engineer to require an exact duplicate. If product quality is comparable, the only remaining test recognized by the courts is whether or not it will do the same job. Of course, you may have to wait until you get to court before being granted relief on this point, so give serious consideration to negotiating a compromise settlement on smaller claims. Due consideration should be given to the fact that another valid reason for rejecting a proposed "or-equal" product is the fact that it may not fit the physical dimensions of the building space shown in the contract documents.

Remember also that the public agency (owner) or its architect-engineer cannot provide any prebid information about the acceptability of any proposed substitute products, so if a contractor plans to propose a product that is different from one of the brands named in the specifications, it does so at the risk of its being ruled unacceptable (not equal) by the architect-engineer, and the contractor may find itself forced to furnish either one of the named brands or submit several other proposed product substitutes until one meets the job requirements. Furthermore, a contractor cannot claim any added compensation for making the change, regardless of the fact that its bid may have been based upon a lower-cost product. That is the risk of basing a bid on an "or-equal" product.

Many contractors have been heard complaining to public agencies about their opinions regarding the specification of "or-equal" products. However, it should be borne in mind that an agency's hands are just as tied to these requirements as are those of the contractor. It is not an issue for contractual negotiations, but purely a legislative matter. To change the law, it is necessary to appeal to legislators, not to the agency with whom the contractor must do business.

Typically, in many private construction projects, contractors are accustomed to making substitutions of products somewhat informally; occasionally, without even the owner's or architect-engineer's knowledge or consent. This is a risky decision for a contractor to make in any case, but in a public works project it is simply not tolerated. Failure to provide the exact product called for in the specifications can lead to cost overruns of the contractor's budget when, after all of the delays and negotiations, the removal and replacement of nonacceptable products have been completed. In all cases, the rules of the game call for a formalized procedure for the submittal of proposed substitute "or-equal" products for consideration and failure to comply with this regulation will cost the contractor in both lost time and cash. Although the means of providing acceptable substitute products certainly do exist, with the contractor pocketing the savings, such must be done in strict accordance with the terms of the contract documents.

Protection for Subcontractors

In some jurisdictions, the bidder must submit the names of proposed subcontractors along with its bids and is prohibited by law from changing the previously listed subcontractors after the award. A word of caution is in order. In some areas of the country, these requirements must be complied with to the letter; otherwise, the contractor may suffer financial losses because of the penalties or forfeitures for noncompliance. The purpose of such laws is to prevent bid shopping by the general contractor who, after successfully obtaining a contract, starts shopping around for a better price than it obtained before the bids were opened. The financial advantage to the general contractor is obvious; however, the adverse impact on subcontractors is substantial. And this is what prompted subcontractors to lobby for legislative protection from unscrupulous general contractors.

Public Advertising of a Project

The advertising of a proposed project in a newspaper of general circulation is generally a prerequisite to the receiving of bids on all public works projects except federal projects, which are prohibited from advertising in any medium where they must pay for the advertising. Federal projects are advertised in the Commerce Business Daily (CBD) published by the U.S. Government Superintendent of Documents. Advertisements will specify where the bids are to be received and specify a precise due date including the time of day. By law the public agency is prohibited from accepting any bids after the stipulated time. Thus, even if you rush into a bid opening two minutes after the specified time, the agency has the right to return your bid to you unopened. There is some precedent for the acceptance of a late bid that has been mailed; however, this is so only if the postmark is at least five days prior to the designated date of opening bids. Some precautions are advised for any contractor planning to mail in its bid. Do *not* use postage meter stamps, as the date on the metered stamp is not acceptable evidence of the mailing date. Only a post-office circular hand-stamped cancellation or a certificate of mailing is acceptable evidence. The best course of action is to mail the bid at the post office and ask for a certificate of mailing.

Bonding Requirements

Bonding and bonding capacity are the life's blood of a contractor doing construction work in the public sector, as nearly every public works contract

CHAPTER 2 PUBLIC WORKS VS. PRIVATE CONTRACTS

in every jurisdiction over a certain nominal amount requires that the contractor post a Performance Bond and a Payment Bond in an amount specified by applicable law. In some jurisdictions, the two bonds are combined into a single bond. Simply stated, without bonding capacity, it is virtually impossible for a contractor to work as a general contractor on a public works project.

Because bonding is so important to a public works contractor, it is necessary for a contractor to guard zealously its bonding capacity and strive to improve it at every opportunity. A good relationship between a bonding company and a contractor doing public works construction is an absolute necessity. The bonding company and bonding agent are the contractor's allies and should be treated as such.

In general, a bonding company will issue a contractor a bond only after it is satisfied that the contractor is a viable and ongoing entity that represents a very low risk in terms of default on the bond. The amount of bonding capacity that a contractor has at any time is normally determined by the bonding company based upon the contractor's short-term assets, less certain contingent liabilities.

Bonding capacity normally varies within the range of one to ten times the contractor's short-term assets. For example, if the contractor has $500,000 in net quick assets or working capital and we assume that the bonding company is inclined favorably toward the contractor, using a ten-to-one multiple, the bonding company would write a maximum of a $5 million bond for that contractor. Should the contractor have outstanding contingent liabilities alleged against it, the amount of that liability may be deducted from its total bonding capacity. There are no absolute rules with regard to bonding capacity, as a great deal of discretion resides in the bonding company agent and the surety company when writing bonds.

In light of the bonding company's importance to the contractor on public works projects, it is important to treat the bonding company as an ally and to keep it well informed in advance of any and all problems as they arise during the project.

For a contractor to obtain public works construction contracts, it must be able to post a Bid Bond and have sufficient bonding capacity to handle the entire job. The cost of the Performance Bond and the Payment Bond must be considered in figuring the bid. Generally, the face amount of each of these bonds is equal to 100 percent of the total construction cost. In order to obtain bonding, a contractor must have a good relationship with its bonding company.

PRINCIPAL DIFFERENCES

Relief from Bid Errors

Occasionally, a contractor will make a mistake in its bid. Sometimes, this will result in the pricing of a project well under what it does cost to perform the work. If under such circumstances you turn out to be the low bidder, you may find yourself in a difficult situation. The basic rule followed both in federal procurement and by those states that offer statutory relief to a bidder who has made such a mistake is that relief is given where the mistake is an error of fact, but not where it is the result of an error of *judgment*. If you need to be relieved of a bid, be prepared to show that a mistake was made due to an error of fact.

An example of an error of fact would be a mistake resulting from a clerical error such as incorrect multiplication or addition. While you may be called upon to show evidence from your backup calculations that your intent is consistent with the statement of error, such errors are generally held to be excusable and the bidder may be allowed to withdraw its bid and have its bid bond or other bid security returned. Such an error of fact is often held to be excusable but *not correctable,* which has the effect of barring you from resubmittal with corrected data. In some states, once a contractor has been given relief, it is permanently barred from ever rebidding the same project.

Errors that have been determined to be errors of judgment are generally neither excusable nor correctable, and the bidder may well find itself contractually bound to a contract that is very difficult to complete without incurring financial loss. At that point, the bidder may choose to refuse to enter into a contract and most certainly forfeit its bid bond or other security; and in some cases it may face litigation by the owner to force the contractor to perform the contract at the stated price, even though erroneous. In any event, the contractor is certainly going to be in trouble with its bonding company and will face the efforts of the bonding company to recover any losses it suffered on its bonds. However, in some cases, forfeiture of the bid bond may be the cheapest way out for the contractor.

Documentation

The formality of public contracts cannot be overemphasized. The documentation requirements and paper shuffling should not be considered superfluous. In this day and age, a public agency is not only buying the project that you are called upon to construct, but it is also contracting with you to provide a paper trail throughout the entire project. This paperwork is every bit as important to politicians as the project itself, as in the future an elected official's entire career may hinge on his or her ability to produce the documents covering every phase of the project. The answer is simple. Instead of complaining how it adds to the construction cost and slows down your operations, play the game by the rules and see to it that the cost of the added paper chase is included in the bid that you submit. One of the most common errors noted in bids of contractors who are new to construction in the public sector is their failure to recognize the added cost of doing business with the public and to include such costs in their bids.

CONTRACTUAL STATUS OF SUBCONTRACTORS AND SUPPLIERS

Unless there are multiple prime contractors, a construction contract for public works is normally entered into between the public agency and a single general contractor. Although the general contractor may have been required to list its subcontractors, as is the requirement in certain jurisdictions, there is no contractual relationship between the subcontractors and the public agency. Furthermore, there is no direct contractual relationship between the public agency and the material suppliers to the contractor either.

It is normal contracting practice, both in the private and the public sector, for the general contractor's subcontracts to incorporate the terms and conditions set forth in the general contractor's contract with the public agency. Despite that incorporation, the subcontractors are not in privity of contract with the owner—that is to say they do not have a contractual relationship with the owner—and must seek payment for their work through the general contractor. However, some states have enacted special statutes that allow subcontractors and material suppliers to directly file actions against the public entity that owns the project for payment due the subcontractor or supplier from the general contractor using a legal device known as a "Stop Notice."

Without specialized statutory protection, however, subcontractors and suppliers must seek redress from the general contractor and any other parties on the basis of applicable contract law.

CHAPTER 3

Risk Allocation

- Risk and Liability Sharing
- Definition of Risk
- Scope and Applicability to Construction
- Identification and Nature of Construction Risks
- Shifting of Risk by Contract
- Who Should Accept What Risks?
- Minimizing Risks and Mitigating Damages
- Types of Risks and Allocation of Those Risks
- Contractual Allocation of Risk
- Risks Reserved to the Contractor
- Contractor Participation in Value Engineering
- Disputes

CHAPTER 3 RISK ALLOCATION

In recent years, a great amount of lip service has been paid to the concept of risk management through the medium of risk allocation and liability sharing. This seems to be the inevitable result of the many losses suffered by contractors, owners, and architect-engineers alike in connection with the projects they build. In general, contractors were always expected to bear the responsibility for as many construction risks as the agencies could pass off on them through the indiscriminate use of exculpatory clauses for everything from risks of unforeseen underground conditions to substandard designs and specifications.

Legislation, case law, and public policy have now changed the situation. By retaining responsibility for some risks, the public agencies have attempted to reduce project costs that courts often forced upon them, regardless of contract language.

Contractors are beginning to realize that they have the means and the right to recover for losses that are the result of the imposition of unfair conditions.

Contractors no longer have to sit idly by and suffer all of the risks and accompanying financial losses while being locked into a guaranteed maximum price. Favorable court decisions prove that contractors can fight back. Far from trying to operate in a risk-free environment, contractors understand that risk is a part of their business. All that is wanted is fair reimbursement for taking such risks.

Although part of the job of the contractor's project manager or superintendent is to minimize exposure of the contractor to risk of losses during construction, the contractor should not assume other risks that are rightfully within the responsibility area of the public agency. Some risks may be transferred to the contractor by contract. However, it should be recognized that all risks are rightfully the agency's unless transferred or assumed by the contractor or an insurance carrier for fair compensation. Theoretically, the principal guideline in determining which party should bear a risk is to determine which party has both the competence to assess the risk and the expertise necessary to control or minimize it. The choice must be made before allocation can take place. One approach to the decision-making process is outlined in the flowchart in Figure 3.1.

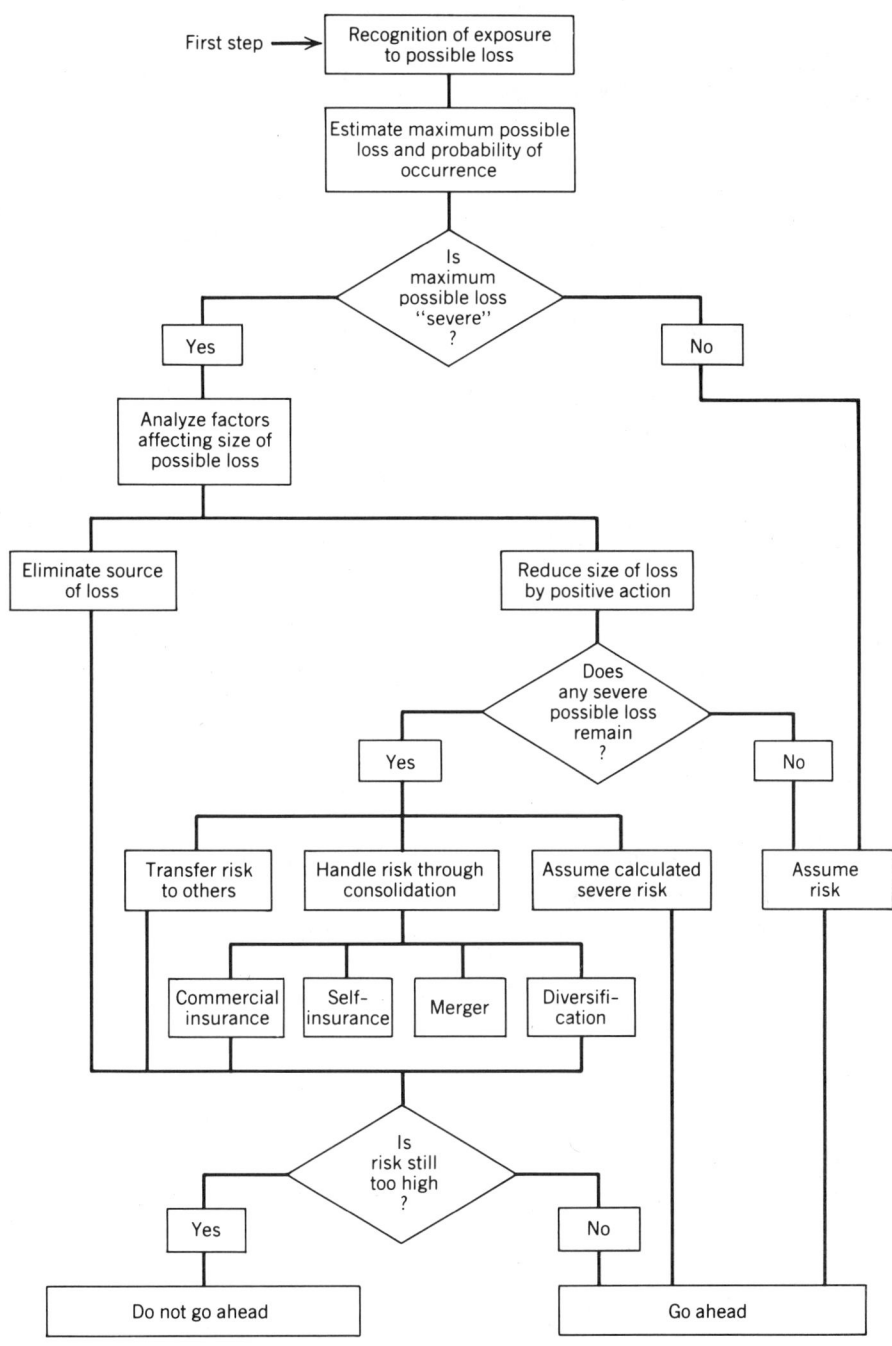

FIGURE 3.1 Logic flowchart for risk decisions.

CHAPTER 3   RISK ALLOCATION

RISK AND LIABILITY SHARING

The first fact that must be recognized is that risks, especially in construction, do exist and are not necessarily fairly distributed. Some kinds of risks must be recognized as inevitable in engineering and construction and therefore philosophically and realistically must be accepted as part of the situations to be dealt with. Managing risks, as the term is used here, means minimizing, covering, and sharing of risks—not merely passing them off to another party.

Although some risks can be avoided, risk management and liability sharing deal primarily with the following concepts:

1   Minimizing risks, regardless of whose risk it is.
2   Equitable sharing of risks among the project participants.

The parties involved must be able to sit down together, prior to the start of the work, to come to an understanding of the realities of the risk responsibility, assumption, and allocation. The parties must be prepared to discuss and decide on the following issues:

1   What levels of risk are realistic to assume.
2   Who can best assume each risk.
3   What levels and kinds of risk are properly and most economically passed on to insurance carriers.

Risk exists wherever the future is unknown. Because the adverse effects of risk have plagued people since the beginning of time, individuals, groups, and societies have developed various methods for the management of risk. Because no one knows the future exactly, everyone is a risk manager, not by choice but by sheer necessity.

DEFINITION OF RISK

Risk has been defined in various ways. There is no single "correct" definition. To emphasize the major objective of risk management, we will define risk as the variation in the possible outcome that exists in nature for a given situation. Another way to clarify this definition of risk is to distinguish between risk and probability. Risk is a property of an entire probability distribution, whereas there is a separate probability for each outcome.

Both risk and probability have their objective and subjective interpretations. The true state of things may be different from the way it appears. Because a person acts on the basis of what is believed to be correct, it is important to recognize this distinction. To the extent that a person's estimates are incorrect, that person's decisions are based upon false premises. Consequently, risk managers must constantly strive to improve their estimates. Even with perfect estimates, decision making about risk is a difficult task. Uncertainty is the doubt that a person has concerning his or her ability to predict which of the many possible outcomes will actually occur. In other words, it is a person's conscious awareness of the risk in a given situation.

SCOPE AND APPLICABILITY TO CONSTRUCTION

Risk, as such, is present in all situations and businesses. As applied to construction, the principal categories include

1. Construction-related risks.
2. Physical risks (subsurface conditions).
3. Contractual and legal risks.
4. Performance risks.
5. Economic risks.
6. Political and public risks.

As it may well be imagined, not all of the items are considered in the same order of priority by all contractors, public agencies, and architect-engineers, and there are also considerable differences of viewpoint as to the percentage of each item that should be shared by the various parties.

As a means of defining the range or spectrum of risk and liability sharing in construction, consider that at one end of the spectrum the contractor is assigned the entire risk or liability with no risk to the public agency. This usually results in a high bid price; in effect, the contractor is acting as the insurer of the agency, and its bid is expected to increase accordingly.

At the other end of the spectrum, the contractor is released from all risks and liability, and the burden is assumed in its entirety by the public

CHAPTER 3 RISK ALLOCATION

agency. This could result in a lower cost; in effect, the agency would have assumed a self-insuring role.

Either end of the spectrum has both advantages and disadvantages. The public agencies must thoroughly assess their own situation to determine what allocation or distribution of risks would serve them best. For instance, a large city, with an excellent engineering force of its own, might find it to its advantage to minimize contractor risk and assume more of the risk on behalf of the large amount of engineering resources and a thorough and expert design staff could afford to assume a larger portion of the risk to obtain the overall benefit of lower costs. Other factors that should influence the agency's assessment are the type of construction (does it include high-risk construction such as underground work?); the degree of detail; accuracy and/or completeness of the plans and specifications; in some cases, the urgency of the project; legal requirements dictated by the specific type of location at a project and/or requirements of the project's source of funding; and other similar considerations.

IDENTIFICATION AND NATURE OF CONSTRUCTION RISKS

Although construction risks can be categorized in many ways, only four groups are presented here—namely, Physical, Capability, Economic, and Political & Societal. In the process of identifying risks, only those that are created by the parties themselves in their attempts to transfer risks are included.

SHIFTING OF RISK BY CONTRACT

There are two basic precepts or guidelines that should be recognized as criteria for the sharing of risks inherent in a construction project.

1. Who can best control the risk?
2. Who can best foresee the risk?

The courts, however, add two additional guidelines of their own:

1. Who can best afford the risk ("deep pockets" principle)?
2. Who most benefits or suffers if the risk materializes?

The principal means available for the contractual allocation or "reallocation" of risk are the construction specifications for the construction contract and the owner/architect-engineer agreement for the design of the project. Under the format endorsed by the Engineers Joint Documents Committee,[1] which has been approved and endorsed by The Associated General Contractors of America, such provisions would logically be spelled out in the General Conditions or Supplementary General Conditions of a contract. In the case of a public works contract, the contractor does not have the opportunity to participate in the wording of the agreement between the parties. However, in private contracts, this issue could become a valid bargaining point. The wording should in any case be prepared by competent legal counsel, avoiding the attitude adopted by some attorneys that a client is best served by exculpatory clauses that would seem to relieve a client of any responsibility for anything, including the negligence of its own personnel. The weakness of this outlook may become more evident during the settlement of some of the large claims that frequently follow such issues.

Exculpatory Clauses

An exculpatory clause is one that attempts, by specific language, to shift the risk or burden of risk from one party to another. As the impact of such clauses can be very great, a contractor must be very careful to review the contract for such exculpatory clauses in order to determine what risks are being shifted to it and how the bid should be adjusted to reflect that risk.

A contract, for example, may contain an exculpatory clause that is intended to shift the responsibility for the engineer's errors and omissions to the contractor, such as illustrated in the following example:

> If the Contractor, in the course of the work, becomes aware of any claimed errors or omissions in the Contract Documents, it shall immediately inform the Engineer. The Engineer shall then promptly review the matter and if an error or omission is found, the Engineer will advise the Contractor accordingly. After discovery of an error or omission by the Contractor, any related work performed by the Contractor shall be done at its own risk unless otherwise authorized in writing by the Engineer.

---

[1] *Standard Forms of Agreement* and Engineer-, Owner-, and Construction-Related Documents prepared by the Engineer's Joint Contract Documents Committee (American Consulting Engineers Council; American Society of Civil Engineers; Construction Specifications Institute; and National Society of Professional Engineers).

CHAPTER 3 RISK ALLOCATION

Under such a provision, the public agency may claim that work done by the contractor in accordance with incorrect plans or specifications will not be paid for. Although it is the intent of the clause to prevent the contractor from knowingly exploiting any errors or omissions the contractor may have become aware of to the detriment of the agency, the contractor effectively becomes an insurer of the architect-engineer's work. Such a clause shifts a risk that can substantially increase risk to the contractor.

In the following example of an exculpatory clause, a similar situation is created.

> The Contractor shall give all notices required by law and shall comply with all laws, ordinances, rules, and regulations pertaining to the conduct of the work. The Contractor shall be liable for all violations of the law in connection with work provided by the Contractor.

Under the terms of the provision just presented, the contractor can be held liable for violations of the law, *even though those violations were the result of building the project in accordance with the plans and specifications that violated the law*. While it is clearly the responsibility of the design professional to prepare plans and specifications that comply with all applicable laws, this clause unfairly shifts the responsibility for such compliance to the contractor.

Although most exculpatory clauses have been interpreted by both state and federal courts in light of what is equitable to all the parties involved, some courts have ignored equitable considerations and considered only the strict, often harsh terms of the clause, regardless of harshness. In a recent New York court decision, Kalisch–Jarcho, Inc. was a successful bidder on an eight million-dollar heating and air conditioning contract for the construction of a new police headquarters in New York City. The contract included a provision in which Kalisch–Jarcho agreed to make no claim for damages for delay in the performance of the contract occasioned by an act or omission of the City, and any such claim would be compensated by an extension of time only. Kalisch–Jarcho sued New York City for breach of contract, claiming damages for 28 months of delay. The trial court awarded Kalisch–Jarcho approximately one million dollars, but the decision was reversed upon appeal. The court found that the exculpatory clause protected the City from a claim of damage as there was no evidence that the delay had been intentional. [*Kalisch–Jarcho, Inc. v. City of New York,* 448 N.E. 2d 413 (N.Y. 1983)]. Surely, Kalisch–Jarcho never anticipated that the City would cause a 28-month delay; nonetheless, under the terms of the contract, Kalisch–Jarcho assumed that risk.

WHO SHOULD ACCEPT WHAT RISKS?

There is no fixed rule to answer this question, but the chart in Figure 3.2 suggests a starting position for theoretically determining who should bear what risks. It should be recognized, however, that in some cases more than one party to the contract may share a common risk. In such cases, though, the risk may be shared in name only, as the specific risk carried by each party may differ materially in terms of the specific details of the risk they carry.

MINIMIZING RISKS AND MITIGATING DAMAGES

The contractual provisions and the methods used in the allocation of risk should be clear enough so that all parties understand in advance what risks they are obligated to assume under the contract, how they will be compensated for the risk, and that they will have the means to control or monitor the risk. Otherwise, the contractor may find itself paying for risks that it did not budget for in the cost estimate of the job.

TYPES OF RISKS AND ALLOCATION OF THOSE RISKS

Most contracts specifically address and allocate various risks. The following are various risks and how they are generally allocated.

Site Access

This is an early risk and one that the public agency usually retains. The contractor is not in a position to influence those who are in control of the site to make it available. However, the risks associated with permit requirements that relate to the contractor's capacity or safety control program can and should be rightfully assumed by the contractor.

| CONSTRUCTION RISK ALLOCATION TO PARTICIPANTS ||||||
TYPE OF RISK	CONTRACTOR	OWNER	ENGINEER	COMMENTS
Site access		●		
Subsurface conditions		●		a
Quantity variations	●	●		b
Weather	●			c
Acts of God		●		
Financial failure	●	●	●	
Subcontractor failure	●			
Accidents at site	●			
Defective work	●			
Management incompetence	●	●	●	
Inflation	●	●		d
Economic disasters		●		
Funding		●		
Materials and equipment	●			
Labor problems	●			
Owner-furnished equipment		●		
Delays in the work	●	●	●	e
Environmental controls		●		
Codes and regulations		●		
Safety at site	●			
Public disorder		●		
Union strife	●			
Errors and omissions			●	
Conflicts in documents			●	
Defective design			●	
Shop drawings			●	

[a] Can be transferred to the contractor; however, the owner has the obligation to undertake precontract exploration measures, and the designer has the responsibility to design for the conditions expected.
[b] Contractor can be expected to assume risk up to 15 to 25 percent. When quantities are dependent on unforeseen subsurface conditions, the owner must assume the risk.
[c] Normal weather for the time and location only. Unusual inclement weather that delays work is the owner's responsibility.
[d] Sharing of escalation risk should be limited to 12- to 18-month span.
[e] Usually, the contractor's risk; however, the owner could incur some liability.

FIGURE 3.2 Construction risk allocation to participants.

Subsurface Conditions

The state of soils, geology, or ground water such as that shown in Figure 3.3 that are known at the time of bidding are often transferred to the contractor, who is usually in a better position to assess and account for the impact of these conditions on the project cost and time. However, as an essential part of the transfer process, the agency has the responsibility to undertake pre-contract exploration measures or allow the necessary time for bid submittal so that the contractor can perform its own explorations, and likewise the designer has the responsibility to design for the conditions expected. As for "Differing Site Conditions," it has become a general principle of public contract law that the agency bear such risks. In that way, the bidding contractors do not have to worry about becoming insurers of the project by raising

FIGURE 3.3 Ground water conditions that are made known to the contractor at the time of bidding can be reasonably transferred to the contractor.

their bids to cover all contingencies. To the extent that this is not feasible, the agency should determine the degree to which the agency retains a portion of the risk under the Differing Site Conditions clause.

Weather

Except for extremely abnormal conditions, this is a risk that the contractor may or may not be expected to assume, depending on the type of project, anticipated length of project time, and normal weather conditions at the project locale. Often, the contractor is expected to assume the risk because it is in the best position to assess and account for the impact of the weather.

Acts of God

Events such as a flood, earthquake (Figure 3.4), hurricane, or tornado are exposures that should not be transferred, but retained by the public agency. However, the architect-engineer should reasonably assume the responsibility for designing in such a way as to minimize their impact. Fire, however, to the extent that it is influenced by the contractor's operations, may be one catastrophe shared with the public agency, but it is clearly one of the easiest risks to insure for and is usually a part of the required contractor's insurance coverage paid for by the public agency indirectly.

Quantity Variations

Within reasonable tolerances, quantities of work can be estimated and any variances assumed by the contractor for all quantities in excess of the 15 or 25 percent unit price guarantee limits set under the contract. Where a reliable estimate of quantities is dependent upon subsurface or other unforeseen conditions, significant variations may be shared, but only to the extent that exploratory information is available. Quantity changes caused by the public agency's late changes in its requirements should be at the risk of the public agency. Some types of variation, such as tunneling overbreak, are contractor-controlled and should be accepted by the contractor.

Capability-Related Risks

The contractor, the public agency, and the architect-engineer all have different capacities and expertise. The result of the failure of any one of these parties to live up to its best standards should be borne by the failing party.

TYPES OF RISKS AND ALLOCATION OF THOSE RISKS

FIGURE 3.4 Act of God: earthquake damage.

Unfortunately, this does not always happen. Too often, the contractor who has the task of building a project carries the burden of the public agency's or architect-engineer's failure. This, in turn, results in the contractor's performance task being either unfeasible or feasible only at considerable extra cost to the contractor.

Defective Design

The tremendous expansion of construction has placed greater burdens upon the design professions. Maintaining performance standards in the face of this is quite difficult, and all too often, design or specifications' defects occur that create construction problems. Unfortunately, it is usually the contractor and the public agency that suffer the consequences of such failures instead of the architect-engineer who probably created the problem in the first place. Design failures or constructability errors are becoming more and more apparent, and the architect-engineer should be made to bear the cost of any such failures that are the result of the negligence of the architect-engineer. Often, the ill-advised use of performance specifications are used as an escape from the responsibilities of design.

Subcontractor Failure

This is a risk that is properly assumed by the contractor, except when it arises from one of the other listed risks that is attributed to the public agency or the architect-engineer. The prime or general contractors are in the best position to assess the capacity of their subcontractors, and therefore it is they who should bear the risk of failing to properly assess that risk.

Defective Work of Construction

This should be the contractor's risk, to the extent that the problem is not caused by a design defect.

Accident

These exposures are inherent to the nature of the work and are best assessed by the contractor and its insurance and safety advisor. Furthermore, a contractor has the most control over site conditions that can increase or decrease its accident exposure.

    The safety record on a construction project is so heavily affected by the contractor's methods, site conditions, worker attitudes, and supervisor awareness that the public agency will want the contractor to provide appropriate insurance, with the public agency and often the architect-engineer as also-named insureds.

Managerial Competence

This is a risk that must be shared by each party, as they each have their own set of managers. It is an ongoing challenge for each organization to assign personnel according to their respective competence levels.

Financial Failure

This is a risk not frequently mentioned and can happen to any of the parties to the contract. Although relatively infrequent, the order of magnitude of such failure should be considered. It is a shared risk as the parties each need to look at their respective financial resources and the financial resources of their partners in joint undertakings, as well as that of any other parties to the contract. Although the subcontractors often absorb the greatest risk of financial failure because of the risks of nonpayment or unforeseen conditions, if we consider only the principal participants to the construction contract—that is, the owner, contractor, and architect-engineer—from a simple business point of view the general contractor will of necessity absorb the greatest risk of financial failure. Characteristically, subcontractors are most susceptible to financial failure, yet the general contractor is required to perform regardless. A contractor on a public project must carefully consider the financial stability of its subcontractors when bidding a job, as the legal ramifications of having a financially unstable or ultimately bankrupt subcontractor can be very significant indeed.

Inflation

It is one of the world's realities and has been with us for a long time. Every contractor is conscious of its impact on the cost of constructing a project, particularly when the project runs into long, agency-caused delays. The government experts in finance so far have been unable to predict where the country will be, financially, a few years from now, so it is unfair to expect the contractor to do better than the government experts. The contractor's apprehensions will result in higher cost to the public agency, or unwarranted optimism will result in the contractor's own demise. A default resulting from such a failure will result in even greater costs to the public agency. The sharing of the escalation risk should therefore be limited, if possible, to a short span of time. Longer time spans become matters of speculation. Obviously, the longer the project is projected to take to complete, the greater the risk of inflation, and the more a contractor must consider inflation in its bid.

Economic Disasters

These are periodic economic problems of such magnitude that a contractor could not properly assess either their probability of occurrence or cost impact. An example might be situations such as nationwide strikes, tax-rate changes, and similar large scale incidents. The public agency should retain the risk of such disasters.

Funding

This is obviously a risk that is beyond the capacity of the contractor to control. Improper administration by a public agency may cause delays or create interest costs that are not anticipated and financing problems that cause delays in payments to the contractor. There is no moral justification for a competent contractor being driven out of business by delayed compensation for services rendered. This is especially true in the protracted negotiation of change orders. All too often, the agency plays the cash flow game to lever dispute negotiations to the agency's advantage. Some large contractors with financial capability may be able to fund these delays, with great outlay of interest, but all too often the smaller contractor cannot even survive.

Labor, Materials, and Equipment

These involve considerable risks. The availability and productivity of the resources necessary to construct a project are risks that are proper for the contractor to assume. The expertise of the contractor should allow the assessment of the cost and time required to obtain and apply these resources. This is the basic service that the public agency is paying for.

Acceleration or Suspension of Work

This is a risk properly retained by the agency, but all too often it is pushed onto the contractor in the form of *constructive acceleration* or *constructive suspension*. It is necessary to have an objective appraisal of the facts underlying the situation and acceptance of responsibility where it properly belongs. It is important to realize that this applies to legitimate acceleration, however, and not to false claims of acceleration.

Political and Societal Risk

This is an area of growing importance to any effort at risk allocation. It is an area in which the political and social pressures from parties having little interest in a project but a great impact on such a project greatly influence its outcome. This is an unclear area and deserves much careful thought as to how the risk should be allocated. In some cases it is clear; in others, it is not.

Environmental Risks

Those of the project itself rightfully belong to the public agency alone and should be retained by the agency, except to the extent that they are influenced by construction methods determined by the contractor, or created by suppliers controlled by the contractor.

Regulations

These are the rules made by governmental bodies in the social area, such as safety and economic opportunity, under which the contractor must be expected to operate. Although there is additional risk in this area, it is similar to the "work rules" established under union contract agreements.

Public Disorder and War

These are political catastrophes of such impact that their risk is best retained by the public agency, lest it become necessary to pay an unusually high price for transferring the risk to another party.

Union Problems

These and all that accompanies them are risks that should be taken by the contractor. Unjustified work rules and similar problems are all risks that the contractor must assess and provide for. Very often in public jobs, however, to again minimize bids and limit the contractor's role as an insurer, the risk of labor problems may be taken by the agency insofar as it represents an excusable but noncompensable delay.

CONTRACTUAL ALLOCATION OF RISK

As described earlier in this chapter, there are two basic precepts and guidelines for the sharing of the risks inherent in a construction project.

1. All risks are rightfully those of the public agency unless and until contractually transferred to or assumed by the contractor or insurance underwriter for a fair compensation.
2. The principal guideline in determining whether a risk should be so transferred is whether the receiving party has both the competence

CHAPTER 3 RISK ALLOCATION

to fairly assess the risk and the expertise necessary to control or minimize it. In addition, it is important to establish whether the shift of the risk from the public agency to another party will result in a savings to the agency and the public.

Certain risks have been referred to as those that the public agency *should* assume. Unfortunately, in the real world, relatively few agencies are far-sighted enough to recognize the merits of doing so, and the agency is the one that sets the terms of the contract. Thus, a contractor frequently will be offered a contract with exculpatory clauses attempting to place *all* of the responsibility on the contractor. Such contracts are offered on a take-it-or-leave-it basis. It may be a long time before many public agencies are enlightened enough to modify their approach to the transference of risk in their construction contracts. Until they do, a contractor would be well advised to refer all such contract documents to its attorney for review and comment before blindly signing an agreement that carries a risk it is unwilling or possibly unable to bear.

To many contractors, risk management is part of the nature of their business. It is one of the tasks they are paid to do. The management of risk first involves a *go* or *no go* decision on risk assumption. To the extent that this process is complicated by unwarranted *risk dumping,* the costs in time and money eventually find their way back to the public agencies in the form of higher prices.

There have been some construction contracts in which total physical risk was assigned to the contractor, including the risk of differing site conditions (sometimes called "changed conditions") and of variations in quantities required to complete the work. Even under such an extreme allocation of risk, the public agency will still retain a very substantial risk that the contractor may not comply with the terms of the contract and properly complete the work within the allotted time. Even if a claim is made under the performance bond, substantial damage may certainly have already occurred. Under the best of circumstances, the agency will still retain material risk. Unreasonable attempts to burden the contractor do not necessarily rid the agency of risk. A contractor can object to an unreasonable allocation of risk by way of a claim which, if won, will return the burden to the agency. Default on the part of the contractor in whole or in part is also a very real risk that an agency can be left with, and there are no winners in that game.

Inclusion in the contract of the frequently used (or misused) disclaimer provision relating to geological information furnished by bidders for underground construction may not be as effective in passing the risk to the contractor as the agency might try to maintain because such clauses are of questionable enforceability. Despite this, such exculpatory clauses continue to be used, and such use actually places the risk on all parties to the contract, without providing relief to anyone. Under such circumstances, a dispute will frequently arise if the geological formation forecasted does not coincide with what is actually found in the field. Such a dispute may not be resolved until many years have passed, and then only at great expense to all of the parties to the contract, with the resulting effect of creating a substantial increase in the cost of the project. In the end, the agency may have to bear a certain risk anyway, by paying a contractor's claim, in spite of any contract clauses to the contrary.

The fact that the contractor carries a substantial burden of risk is beyond dispute. Unfortunately, it seems that the viewpoint of many who design and administer construction contracts for an agency is that the contractor should carry virtually all of the risk, whether provided for in the contract or not. It is precisely this attitude that justifies contractor claims and contributes to the inevitable litigation that follows.

When it may not be apparent how a given risk should be allocated, consideration of what might be the motivation of the parties involved may be productive. Unfortunately, in the traditional construction contracting practice, the public agency allocates almost all of the risks to the contractor, saying in effect, "You deal with all of the construction problems and all the third parties and don't bother me."

This rather one-sided attitude fosters two results.

1 Contractors justifiably add high contingencies to their bids to cover the costs of the risks.
2 Litigation of construction contract claims arise out of the resulting inequities.

Broadly speaking, the public agencies often lose in such actions because the courts frequently reallocate many of the risks that the agencies thought they had laid upon the contractor. Consequently, the agencies sometimes pay for their risks twice—once in bidding contingencies and a second time in court.

Reallocation of Risks to Subcontractors

Many prime contracts include provisions that not only require a contractor to indemnify the agency and its agents for claims and losses, but also to defend and hold them harmless as well. With the ever-increasing costs of

CHAPTER 3 RISK ALLOCATION

defense and construction claims, indemnity provisions in most prime contracts have become more and more onerous for the general contractor. It is not at all uncommon for a prime contract to require the contractor to indemnify, defend, and hold harmless the project engineer or architect, as well as the public agency, from any and all construction claims unless the claim or loss is the result of the *sole* negligence of the architect-engineer or the agency.

If we assume that an agency's sole negligence was the cause of a loss or claim, the contractor and its insurance company would still have to provide the agency with a defense until the fact of sole negligence was proven. In the meantime, the contractor and its insurance company are responsible for the cost of that defense.

One way for the contractor to lessen its indemnity burden is to transfer the indemnity obligation to its subcontractors and their insurance carriers. Generally, the contractor should require its subcontractors to indemnify the public agency as the contractor is required to do under the prime contract, and to indemnify the contractor as well. By drafting its subcontracts to contain indemnity language that reflects at least what is contained in the prime contract, the subcontractors will become obligated to indemnify, defend, and hold harmless the agency, the architect-engineer, and the general contractor as well. However, it should be kept in mind that many states have legislation precluding the shifting of an indemnity burden and obligation. The status of legislation varies from state to state, and such exculpatory clauses may be void in many states.

Reallocating the obligation to indemnify to the subcontractors increases the number of parties potentially liable for any loss or claim and thereby spreads more fully the risk of the project. As a byproduct, it also increases the amount of insurance available to cover any loss or claim.

RISKS RESERVED TO THE CONTRACTOR

In addition to the other types of risks previously referred to, the following are typical of the risks reserved to the contractor:

1 The contractor should bear all risks over which it can exercise reasonable control. These include all matters relating to the selection of construction methods, equipment, and prosecution of the work, except to the extent that this control is affected by the action of third parties.

2. In the area of third-party effects, risks should be allocated to those best able to deal with the third party. This principle would assign to the public agency the risks related to certain government agency regulations and to agreements with adjacent property owners. Risks associated with labor and subcontractor agreements and disputes should be assigned to the contractor.
3. Construction safety should be the responsibility of the contractor, although financial risk with regard to third parties is properly allocated to insurers (either the contractor's or the public agency's), usually the contractor's.

Construction is a highly complex business. Guidelines, recommendations, regulations, contracts, and even legal rulings can only provide direction for judging a particular situation. Some public agencies and some architect-engineers have earned such bad reputations that reputable contractors will not bid on their projects. Others have reputations that even attract bidders who would pass up similar work in other jurisdictions. Conversely, some contractors have earned reputations that invite contract administration "by the book," while others enjoy the ability to secure many contract modifications by negotiation. The risk of an unfavorable reputation (or the benefit of a favorable one) is earned by all parties over a longer period. It is not allocable and is not rapidly changed.

CONTRACTOR PARTICIPATION IN VALUE ENGINEERING

Value engineering by the contractor is a subject that is viewed as controversial by some, but the concept has potential benefits as a means of minimizing risks and cost overruns. Value engineering by the contractor (Chapter 11) involves contractor proposals for changing construction methods or designs as a means of reducing project construction costs without compromising the design principles. Generally, such concepts involve a cost-sharing provision on any savings realized [Federal Acquisition Regulations 48CFR48.101 to 48.104; c.f. CA Public Contracts Code §7101]. Many designers resist such potential intrusions of the contractor into their hallowed ground. Unfortunately, they often feel that their design approach is the only proper one, and it is viewed as a reflection upon their design ability if the contractor dares to question their design. However, the inclusion of such a provision goes a long way to develop a cooperative working relationship between all of the

CHAPTER 3 RISK ALLOCATION

parties that can only result in benefits. As a good contractor, you should be given the opportunity to offer improvements to the project when possible.

DISPUTES

Even in the area of disputes, there is still some leeway for cost-saving measures. After a difference of opinion has been expressed, adequate actions for the resolution of such problems must be taken immediately.

The *second worst* way to handle claims is to ignore them; the *worst* way, however, is to allow them to go to litigation. Most disputes, if handled promptly and vigorously, can be resolved without degenerating into large problems affecting not only the contractor's cost to the project, but possible delays and the assessment of liquidated damages. It is important to both the efficient progress of the work and the lowest cost to both the contractor and the public agency in the performance of Differing Site Conditions work that such changes be negotiated and settled as soon as possible.

CHAPTER · 4 ·

Specifications and Drawings

What Is a Specification? •
Conflicts between Drawings and Specifications •
Content and Component Parts of a Specification •
What Do the Specifications Mean to the Contractor? •
CSI Specifications Format: Its Meaning and Importance •
State Highway Department Formats •
Nonstandard Construction Specification Formats in Use •
Project Specifications vs. Special Provisions Concept •
General Conditions of the Construction Contract •
Supplementary General Conditions •
Technical Provisions of the Specifications •
Addenda to the Specifications •
Standard Specifications •
Special Material and Product Standards •
Access to Special Standards by the Contractor •
Building Codes, Regulations, Ordinances, and Permits •
Types of Drawings Comprising the Construction Contract •

CHAPTER 4 SPECIFICATIONS AND DRAWINGS

WHAT IS A SPECIFICATION?

The Contract Documents define the requirements and geometry of a project that is to be built; the specifications are a part of these documents. The dictionary defines specifications as "a detailed description of requirements, dimensions, materials, etc., as a part of a proposed building, machine, bridge, etc.," and further as "the act of making specific."

The role of the drawings is to define the geometry of a project, including dimensions, form, and details. The specifications are intended to complement this by defining the nature of the materials that are to be used and the description of the workmanship and procedures to be followed in constructing the project.

All too often a contractor, just like many tradesmen, expects the drawings to provide *all* the information required, incorrectly assuming that the specifications are only needed by lawyers in case of dispute. To be sure, the specifications may be needed in case of a dispute, but if properly used and referred to throughout the construction work, they can serve to minimize disputes. Even more important to the contractor is the fact that the specifications are the only document that will spell out the obligations for the administration of the project during its construction. By far the majority of the administrative tasks that the public works contractor will be required to perform are covered in the specific terms of the General Conditions of the contract and nowhere else. Even years of past experience cannot serve as a substitute, and although public works construction is closely regulated by law, the rules of the game still seem to change from project to project. What might have been proper on a previous job may be incorrect on the next.

CONFLICTS BETWEEN DRAWINGS AND SPECIFICATIONS

It should be mentioned here that neglecting the specifications can lead to serious problems. In case something is shown or noted one way on the drawings and described differently *in the specifications*, which will govern? The answer to this question is easy. The specifications will normally take precedence unless it says otherwise in the specifications that the plans will govern. Thus, it is still the specifications that set the controlling criteria. Normally, it is easy to determine the relative importance of one document over another, as most specifications specify the relative order of importance of the different parts of the contract documents in the General Conditions of the construction contract. However, it should be of interest that in the absence of such a specific provision, the courts have repeatedly held that the provisions of the specifications will take precedence over the drawings in case of a conflict between the two. [*Appeal of Florida Builders, Inc.,* ASBCA No. 9013, 69-2 BCA 8014 (1969)].

Therefore, if the specifications are the most important single document, a contractor can hardly construct the work in a competent manner unless thoroughly familiar with both the specifications and the construction drawings.

In some cases, the same data are covered in both the drawings and the specifications; this is not the best arrangement, but it happens frequently. The problem here is that often one document is changed during design and the other is overlooked. Unfortunately, when such a problem exists, it is usually because the drawings were updated to reflect the latest changes or corrections, and the specifications may be outdated and incorrect. The basic philosophy still rules, however, and the contractor can only bid the job in accordance with the controlling contract documents. Once the contractor has been awarded the contract, such conflicts can be reviewed, and change orders issued if necessary. Once a conflict is discovered, the contractor is obligated to notify the agency's representative before continuing. If a contractor has already built a portion of a project in accordance with specifications that are subsequently changed, in all likelihood the contractor will be able to successfully claim extra compensation for such additional work.

It should be remembered that some items will appear only in the specifications and not on the drawings; others will appear only on the drawings and not be mentioned in the specifications. This is not necessarily an oversight, or is it to be considered a flaw in the specifications or drawings. Many architectural and engineering firms as a matter of policy prefer not to repeat data on both documents. This is done intentionally as a means of preventing conflicts due to late changes that may be made on one document alone and not the other. The contractor should make certain that its foremen at the project site use the specifications also, as it will minimize construction problems and conflicts.

CONTENT AND COMPONENT PARTS OF A SPECIFICATION

Content of the Specifications

In addition to the well-known *technical* provisions contained in the specifications, it should be clear that the term *specifications* is *not necessarily* limited to the technical portions alone, although advocates of the Construction Specifications Institute (CSI) policies strongly oppose this concept, and they call their document the *Project Manual*. This is common in architectural work, but generally, in public agency contracts, everything that is bound into the specifications document is referred to as the *specifications*; this may include the notice of invitation to bid on the project; the bidding documents and forms, including performance and payment bonds when required and non-collusion affidavits when required; the conditions of the contract, often referred to simply as the "boilerplate" because it provides a protective shield around the contract by anticipating most of the areas of discussion or dispute that might arise, and provides for an orderly method for resolving each such case; and finally, the technical provisions. If the term *Contract Documents* is used, then it legitimately includes all requirements of the contract, including the drawings, and sometimes a book of *standard specifications* by reference as well. Usually, some of the boilerplate documents will specify a list of all items that are to be officially classed as part of the Contract Documents.

Another often misunderstood characteristic of a set of specifications is that the size of a specification is in no way directly related to either the size or cost of a project, but is actually more influenced by how many different trades or materials are involved in the work. Thus, a public restroom building in a park with only one room and a single set of plumbing fixtures may require as thick a set of specifications as a two-story building. But, a highway construction job costing ten times the price of either of the two described buildings may involve a specification of only three or four sections and possibly as few as eight or ten pages of technical provisions supplementing the standard specifications.

Most people in the construction business have, at some time or another, questioned the wisdom of the specification writer. If it will give the contractor any peace of mind, the authors readily agree that not all specifications writers are necessarily knowledgeable in some of the subjects about which they write (sometimes an understatement). In a recent ASCE (American Society of Civil Engineers) questionnaire circulated nationally by the *National Task Committee on Specifications* to engineers, contractors, public agencies, owner-developers, suppliers, and attorneys, the general response from the contractors was that a specifications writer should have field construction experience before becoming a specifications writer. Every contractor has undoubtedly run into specifications in which it seemed obvious that the specifications writer did not have field experience. Part of the problem lies in the procedures often used by architects and engineers in selecting personnel and budgeting time and costs for the production of specifications. All the

CONTENT AND COMPONENT PARTS OF A SPECIFICATION

standardized specification formats in the world cannot cure this problem. If each of the items specified is properly covered, the contractor will have the contractual tools necessary to assure the public agency of quality construction. If these important considerations are neglected, the contractor is deprived of one of the primary tools of the construction business. The result will be a project with many disputes, change orders, and claims. The owner should be made to bear the burden of any deficient drawings and specifications.

Component Parts of a Specification

Generally, most specifications can be divided into three main elements or parts. Although these parts are not necessarily arranged in the same order on each job the contractor will encounter, the various architects and engineers or public agencies responsible will generally keep the content of the specifications within the classifications shown in Figure 4.1.

In addition to the classifications indicated in Figure 4.1, public-funded projects require conformance to a list of minimum-wage rates. If a public project has federal funding and is subject to state labor code requirements, then in addition to the requirements for state wage rates, a complete copy of the applicable federal wage rates must also be bound into the specification, and the contractor is obligated to pay the higher of the two rates if there is a difference. The federal wage rates are normally obtained directly from the federal agency providing the funding for the project. The federal wage rate listings are normally required to be the rates that prevail within ten days of the date of advertising the project for bids.

The Instructions to Bidders is usually a preprinted document and is normally considered one of the contract documents in public works contracts. Thus, the provisions of the Instructions to Bidders are as binding on the bidder and contractor as those of the technical specifications. Failure to comply with its terms can render a contractor's bid *nonresponsive* or *informal,* which then may be used as the justification for rejecting it. The general subject area covered by an Instructions to Bidders document usually includes the following:

Form of bid and signature.
Interpretation of drawings and specifications.
Preparation of the proposal.
List of documents to be submitted with the bid.

CHAPTER 4 SPECIFICATIONS AND DRAWINGS

PART I—BIDDING AND CONTRACTUAL DOCUMENTS

    Notice Inviting Bids   (mostly public works)
    Instructions to Bidders
    Proposal (or Bid) Forms
        Proposal
        Bid Sheets
        Contractor Certificates   (mostly public works)
        List of Subcontractors   (Calif. public works)
        Bid Bond Form   (mostly public works)
        Non-Collusion Affidavits   (mostly public works)
    Agreement and Bonds
        Agreement (contract)
        Performance Bond Form   (mostly public works)
        Payment Bond Form (labor & materials)   (mostly public works)

PART II—CONDITIONS OF THE CONTRACT

    General Conditions (the "boilerplate")
    Supplementary General Conditions (special for the project)
    (In some specifications the supplementary general
    conditions are referred to as "Special Conditions")

PART III—DETAIL (TECHNICAL) SPECIFICATIONS

    (From this point on, the architect/engineer provides
    technical sections covering the various parts of the
    project.)

FIGURE 4.1 Three-part specifications format.

    Bonding requirements.

    What is expected of the successful bidder.

    Insurance policies required.

    Basis for selection of the successful bidder.

The General Conditions or boilerplate as it is often called is one of the most important documents in the specifications to the contractor. It is this document that establishes the ground rules for the administration of the construction phase of the project. The subject matter generally covered in

CONTENT AND COMPONENT PARTS OF A SPECIFICATION

General Conditions is fairly consistent from job to job wherever "standard" preprinted documents of a governmental agency or those of AIA or the EJCDC (Engineer's Joint Contract Documents Committee) or similar organizations are used. Whenever an architect, engineer, or public agency elects to prepare its own general conditions, it frequently lacks many of the essentials and worse yet, has never been "proved" in court. The general subjects covered in most standard public works General Conditions documents include the following:

1. Legal definitions of terms used in the contract.
2. Correlation and intent of the documents.
3. Time and order of the work.
4. Assignment of the contracts.
5. Subcontracts.
6. Where to serve legal notices.
7. Authority of the architect-engineer.
8. Change orders and extra work.
9. Extensions of time for delays.
10. Right of the owner to terminate the contract.
11. Right of the contractor to terminate the contract.
12. Right of the owner to take over work.
13. Obligations of the contractor.
14. Supervision by the contractor.
15. Handling of claims and protests.
16. Lines, grades, and surveys; who performs and who pays (frequently, a specifier includes provisions for surveys in other sections of the specifications, and when he or she does, it often conflicts with the terms of the General Conditions of the Contract).
17. Defective work or materials.
18. Materials and workmanship.
19. Provisions to allow access to all parts of the work.
20. Inspection and tests; how administered and who pays.
21. Coordination with other contractors at the site or nearby.
22. Suspension of all or part of the work.
23. Liquidated damages for delay.

24 Stop notice procedures.
25 Right of owner to withhold payment.
26 Provisions for public safety.
27 Changed conditions (sometimes called unforeseen conditions).
28 Estimates and progress payments.
29 Final payment and termination of liability.
30 Protection and insurance.
31 Disputes; settlement by arbitration.

Part III of the specifications, as outlined in Figure 4.1, refers to that portion of the specifications a lay person usually thinks of when one speaks of specifications. In this portion of the document are the detailed technical provisions that relate to the installation or construction of the various parts of the work and to the materials used in the work. There are several ways of logically dividing these sections into subject areas so as to lend some sort of order to the final documents. Most of the systems, however, generally group specification sections into trade-related functions as a means of easy grouping. This sometimes prompts the complaint from contractors that the sections do not accurately represent the responsibility areas of the various trades. Actually, there is usually no attempt made by the specifications writer to conform exactly to trade jurisdictions, as they vary significantly from one part of the nation to the other. In fact, in some cases, jurisdictional differences may be evident from one adjacent county to the other within the same state.

It should be recognized that on any project there are usually a few technical requirements that would apply to *all* sections equally. In such cases, it has been found desirable to provide a section of the technical specifications, usually at the front of the technical portion, that may be entitled "General Requirements" and that spells out the various requirements of a *technical* nature which apply generally to the entire project. This section should not be confused with any of the rest of Part II, "Conditions of the Contract"; there, the *General Conditions* and *Supplementary General Conditions* are matters of a legal/contractual nature, *not technical provisions*. A common error made by many general contractors in their dealings with their subcontractors is to hand them a single technical specifications section relating to their trade, without copies of either the General Requirements or the General and Supplementary General Conditions of the Contract. This has often created serious problems in the conduct of the work, as all of the boilerplate sections apply to all of the work of each subcontractor as well.

WHAT DO THE SPECIFICATIONS MEAN TO THE CONTRACTOR?

The specifications, in short, are one of the contractor's vital tools. Without them, a contractor cannot possibly do the job in a competent manner without excessive risk. To be able to use these tools effectively, however, a contractor should have an idea of the relative importance of each of the various component parts of the contract documents. The following is a condensed listing of some of the Contract Document components and their relative importance.

> Agreement governs over specifications.
> Specifications govern over drawings.[3] [*Appeal of Florida Builders, Inc.,* ASBCA No. 9013, 69-2 BCA 8014 (1969)].
> Detail specifications govern over general specifications.

Each month on the larger public works projects, the contractor normally applies for and receives monthly partial payments (progress payments) for work completed thus far (to date). The amount of each payment must be in direct proportion to the amount of work completed during the preceding month. It is the responsibility of the agency's or engineer's resident project representative to check the quantities of such work completed by the contractor and review the monthly payment request of the contractor prior to submitting it to the architect-engineer's or public agency's project manager along with a recommendation for payment, if justified. In the handling of such matters, the terms of the General Conditions must be strictly followed, as the procedures for handling such payment claims must follow an orderly, prearranged plan. There are generally no provisions for allowing terms or creating restrictions that were not written into the original contract.

In short, the entire policy for the administration and conduct of the work at the jobsite is established under the terms and conditions of the General and Supplementary General Conditions of the construction contract. The remaining portions of the specifications more properly relate to quality control or quality assurance functions, which are a subsidiary function of the project administrator.

CHAPTER 4 SPECIFICATIONS AND DRAWINGS

CSI SPECIFICATIONS FORMAT: ITS MEANING AND IMPORTANCE

Briefly mentioned previously was the fact that the technical portions of the specifications were generally structured in whatever manner suited the architect or engineer who prepared them for the public agency. In the years past, this problem was even worse, and the contractor would indeed have to be versatile to be required to work from one type of contract documents on one job and at the same time be constructing another similar project nearby from another set of documents that bore no resemblance to the first.

In recent years, an organization called the Construction Specifications Institute (CSI) has tackled the task of injecting some degree of uniformity and standardization into the general arrangement and method of writing construction specifications. To this end the institute has been enormously successful, although it can be seen that the *format* or arrangement and classification system devised was obviously created by architects and engineers whose experience was limited to the construction of buildings, and the resulting format shows very little sensitivity to the problems of engineers engaged in heavy construction projects. The CSI Format was indeed intended for buildings, but once entrenched nothing seemed to be able to change or even alter the system when it became desirous to extend it to other types of engineered construction as well. In fact, when the system is used for certain types of heavy construction projects, some serious formatting conflicts can arise. In any case, the CSI format did create order where none existed before by setting forth a list of 16 standardized *"Divisions"* that are supposed to work for everything. With a little imagination, it can indeed be adapted to many, though not all, nonbuilding construction projects, even though it may appear to be somewhat cumbersome and awkward in certain types of work.

CSI 16-Division Format

The CSI 16-Division format was adopted by the AGC (Associated General Contractors), the AIA (American Institute of Architects), the NSPE (National Society of Professional Engineers), and others in the United States and Canada in the form of a document entitled "Uniform System for Building Specifications." Note that word "building" again: There is no mention of heavy construction projects. Nevertheless, the document is widely used for both building and some engineering work and is popularly known as the *CSI Format*. This system has been officially adopted for *all* construction work by the U.S. Army Corps of Engineers; the U.S. Navy (NAVFAC); National

CSI SPECIFICATIONS FORMAT: ITS MEANING AND IMPORTANCE

Aeronautics and Space Administration (NASA); the State of New York for public works projects; and by numerous other public and private agencies; however, it has not been adopted nor approved by the American Society of Civil Engineers for heavy construction projects. In fact, that organization had at one time planned to develop an alternate format that would be better suited to heavy construction.

Eventually, all manufacturers followed suit by adopting the CSI classifications for their products, and now most, if not all, building materials are identified with the CSI classification number for filing purposes that corresponds to the CSI division number under which each such product is intended to be grouped. Thus, if you pick up a specification that is in the CSI format, even without a table of contents, you should be able to find the general location of any section you need. For example, you should automatically turn to Division 3 if you are looking for concrete or Division 16 if you are looking for electrical work (Figure 4.2).

From a civil, mechanical, or electrical engineer's standpoint there are still some problems to be solved in devising a uniformly adaptable specification "Division" list. Whenever this fact is recognized and the matter studied objectively from the standpoint of those professionals who construct both buildings as well as heavy engineering works or systems, the industry may

```
 1   General Requirements
 2   Site Work and Utilities (includes all civil work)
 3   Concrete
 4   Masonry
 5   Metals
 6   Wood and Plastics
 7   Thermal and Moisture Protection
 8   Doors and Windows
 9   Finishes
10   Specialties
11   Equipment
12   Furnishings
13   Special Construction
14   Conveying Systems
15   Mechanical
16   Electrical
```

Note that whenever a job does not use a certain Division, it is simply skipped. . .but the numbers of the remaining Divisions still never change.

FIGURE 4.2 List of CSI 16 Divisions.

BIDDING REQUIREMENTS, CONTRACT FORMS, AND CONDITIONS OF THE CONTRACT

00010	PRE-BID INFORMATION
00100	INSTRUCTIONS TO BIDDERS
00200	INFORMATION AVAILABLE TO BIDDERS
00300	BID FORMS
00400	SUPPLEMENTS TO BID FORMS
00500	AGREEMENT FORMS
00600	BONDS AND CERTIFICATES
00700	GENERAL CONDITIONS
00800	SUPPLEMENTARY CONDITIONS
00850	DRAWINGS AND SCHEDULES
00900	ADDENDA AND MODIFICATIONS

Note: Since the items listed above are not specification sections, they are referred to as "Documents" in lieu of "Sections" in the Master List of Section Titles, Numbers, and Broadscope Explanations.

SPECIFICATIONS

DIVISION 1—GENERAL REQUIREMENTS

01010	SUMMARY OF WORK
01020	ALLOWANCES
01025	MEASUREMENT AND PAYMENT
01030	ALTERNATES/ALTERNATIVES
01040	COORDINATION
01050	FIELD ENGINEERING
01060	REGULATORY REQUIREMENTS
01070	ABBREVIATIONS AND SYMBOLS
01080	IDENTIFICATION SYSTEMS
01090	REFERENCE STANDARDS
01100	SPECIAL PROJECT PROCEDURES
01200	PROJECT MEETINGS
01300	SUBMITTALS
01400	QUALITY CONTROL
01500	CONSTRUCTION FACILITIES AND TEMPORARY CONTROLS
01600	MATERIAL AND EQUIPMENT
01650	STARTING OF SYSTEMS/COMMISSIONING
01700	CONTRACT CLOSEOUT
01800	MAINTENANCE

DIVISION 2—SITEWORK

02010	SUBSURFACE INVESTIGATION
02050	DEMOLITION
02100	SITE PREPARATION
02140	DEWATERING
02150	SHORING AND UNDERPINNING
02160	EXCAVATION SUPPORT SYSTEMS
02170	COFFERDAMS
02200	EARTHWORK
02300	TUNNELING
02350	PILES AND CAISSONS
02450	RAILROAD WORK
02480	MARINE WORK
02500	PAVING AND SURFACING
02600	PIPED UTILITY MATERIALS
02660	WATER DISTRIBUTION
02680	FUEL DISTRIBUTION
02700	SEWERAGE AND DRAINAGE
02760	RESTORATION OF UNDERGROUND PIPELINES
02770	PONDS AND RESERVOIRS
02780	POWER AND COMMUNICATIONS
02800	SITE IMPROVEMENTS
02900	LANDSCAPING

DIVISION 3—CONCRETE

03100	CONCRETE FORMWORK
03200	CONCRETE REINFORCEMENT
03250	CONCRETE ACCESSORIES
03300	CAST-IN-PLACE CONCRETE
03370	CONCRETE CURING
03400	PRECAST CONCRETE
03500	CEMENTITIOUS DECKS
03600	GROUT
03700	CONCRETE RESTORATION AND CLEANING
03800	MASS CONCRETE

DIVISION 4—MASONRY

04100	MORTAR
04150	MASONRY ACCESSORIES
04200	UNIT MASONRY
04400	STONE
04500	MASONRY RESTORATION AND CLEANING
04550	REFRACTORIES
04600	CORROSION RESISTANT MASONRY

DIVISION 5—METALS

05010	METAL MATERIALS
05030	METAL FINISHES
05050	METAL FASTENING
05100	STRUCTURAL METAL FRAMING
05200	METAL JOISTS
05300	METAL DECKING
05400	COLD-FORMED METAL FRAMING
05500	METAL FABRICATIONS
05580	SHEET METAL FABRICATIONS
05700	ORNAMENTAL METAL
05800	EXPANSION CONTROL
05900	HYDRAULIC STRUCTURES

DIVISION 6—WOOD AND PLASTICS

06050	FASTENERS AND ADHESIVES
06100	ROUGH CARPENTRY
06130	HEAVY TIMBER CONSTRUCTION
06150	WOOD-METAL SYSTEMS
06170	PREFABRICATED STRUCTURAL WOOD
06200	FINISH CARPENTRY
06300	WOOD TREATMENT
06400	ARCHITECTURAL WOODWORK
06500	PREFABRICATED STRUCTURAL PLASTICS
06600	PLASTIC FABRICATIONS

DIVISION 7—THERMAL AND MOISTURE PROTECTION

07100	WATERPROOFING
07150	DAMPPROOFING
07190	VAPOR AND AIR RETARDERS
07200	INSULATION
07250	FIREPROOFING
07300	SHINGLES AND ROOFING TILES
07400	PREFORMED ROOFING AND CLADDING/SIDING
07500	MEMBRANE ROOFING
07570	TRAFFIC TOPPING
07600	FLASHING AND SHEET METAL
07700	ROOF SPECIALTIES AND ACCESSORIES
07800	SKYLIGHTS
07900	JOINT SEALERS

DIVISION 8—DOORS AND WINDOWS

08100	METAL DOORS AND FRAMES
08200	WOOD AND PLASTIC DOORS
08250	DOOR OPENING ASSEMBLIES
08300	SPECIAL DOORS
08400	ENTRANCES AND STOREFRONTS
08500	METAL WINDOWS
08600	WOOD AND PLASTIC WINDOWS
08650	SPECIAL WINDOWS
08700	HARDWARE
08800	GLAZING
08900	GLAZED CURTAIN WALLS

DIVISION 9—FINISHES

09100	METAL SUPPORT SYSTEMS
09200	LATH AND PLASTER
09230	AGGREGATE COATINGS
09250	GYPSUM BOARD
09300	TILE
09400	TERRAZZO
09500	ACOUSTICAL TREATMENT
09540	SPECIAL SURFACES
09550	WOOD FLOORING
09600	STONE FLOORING
09630	UNIT MASONRY FLOORING
09650	RESILIENT FLOORING
09680	CARPET
09700	SPECIAL FLOORING
09780	FLOOR TREATMENT
09800	SPECIAL COATINGS
09900	PAINTING
09950	WALL COVERINGS

FIGURE 4.3 CSI MASTERFORMAT showing some recommended section titles under each division. (*Source:* Reproduced with permission of the Construction Specifications Institute.)

DIVISION 10—SPECIALTIES

10100	CHALKBOARDS AND TACKBOARDS
10150	COMPARTMENTS AND CUBICLES
10200	LOUVERS AND VENTS
10240	GRILLES AND SCREENS
10250	SERVICE WALL SYSTEMS
10260	WALL AND CORNER GUARDS
10270	ACCESS FLOORING
10280	SPECIALTY MODULES
10290	PEST CONTROL
10300	FIREPLACES AND STOVES
10340	PREFABRICATED EXTERIOR SPECIALTIES
10350	FLAGPOLES
10400	IDENTIFYING DEVICES
10450	PEDESTRIAN CONTROL DEVICES
10500	LOCKERS
10520	FIRE PROTECTION SPECIALTIES
10530	PROTECTIVE COVERS
10550	POSTAL SPECIALTIES
10600	PARTITIONS
10650	OPERABLE PARTITIONS
10670	STORAGE SHELVING
10700	EXTERIOR SUN CONTROL DEVICES
10750	TELEPHONE SPECIALTIES
10800	TOILET AND BATH ACCESSORIES
10880	SCALES
10900	WARDROBE AND CLOSET SPECIALTIES

DIVISION 11—EQUIPMENT

11010	MAINTENANCE EQUIPMENT
11020	SECURITY AND VAULT EQUIPMENT
11030	TELLER AND SERVICE EQUIPMENT
11040	ECCLESIASTICAL EQUIPMENT
11050	LIBRARY EQUIPMENT
11060	THEATER AND STAGE EQUIPMENT
11070	INSTRUMENTAL EQUIPMENT
11080	REGISTRATION EQUIPMENT
11090	CHECKROOM EQUIPMENT
11100	MERCANTILE EQUIPMENT
11110	COMMERCIAL LAUNDRY AND DRY CLEANING EQUIPMENT
11120	VENDING EQUIPMENT
11130	AUDIO-VISUAL EQUIPMENT
11140	SERVICE STATION EQUIPMENT
11150	PARKING CONTROL EQUIPMENT
11160	LOADING DOCK EQUIPMENT
11170	SOLID WASTE HANDLING EQUIPMENT
11190	DETENTION EQUIPMENT
11200	WATER SUPPLY AND TREATMENT EQUIPMENT
11280	HYDRAULIC GATES AND VALVES
11300	FLUID WASTE TREATMENT AND DISPOSAL EQUIPMENT
11400	FOOD SERVICE EQUIPMENT
11450	RESIDENTIAL EQUIPMENT
11460	UNIT KITCHENS
11470	DARKROOM EQUIPMENT
11480	ATHLETIC, RECREATIONAL AND THERAPEUTIC EQUIPMENT
11500	INDUSTRIAL AND PROCESS EQUIPMENT
11600	LABORATORY EQUIPMENT
11650	PLANETARIUM EQUIPMENT
11660	OBSERVATORY EQUIPMENT
11700	MEDICAL EQUIPMENT
11780	MORTUARY EQUIPMENT
11850	NAVIGATION EQUIPMENT

DIVISION 12—FURNISHINGS

12050	FABRICS
12100	ARTWORK
12300	MANUFACTURED CASEWORK
12500	WINDOW TREATMENT
12600	FURNITURE AND ACCESSORIES
12670	RUGS AND MATS
12700	MULTIPLE SEATING
12800	INTERIOR PLANTS AND PLANTERS

DIVISION 13—SPECIAL CONSTRUCTION

13010	AIR SUPPORTED STRUCTURES
13020	INTEGRATED ASSEMBLIES
13030	SPECIAL PURPOSE ROOMS
13080	SOUND, VIBRATION, AND SEISMIC CONTROL
13090	RADIATION PROTECTION
13100	NUCLEAR REACTORS
13120	PRE-ENGINEERED STRUCTURES
13150	POOLS
13160	ICE RINKS
13170	KENNELS AND ANIMAL SHELTERS
13180	SITE CONSTRUCTED INCINERATORS
13200	LIQUID AND GAS STORAGE TANKS
13220	FILTER UNDERDRAINS AND MEDIA
13230	DIGESTION TANK COVERS AND APPURTENANCES
13240	OXYGENATION SYSTEMS
13260	SLUDGE CONDITIONING SYSTEMS
13300	UTILITY CONTROL SYSTEMS
13400	INDUSTRIAL AND PROCESS CONTROL SYSTEMS
13500	RECORDING INSTRUMENTATION
13550	TRANSPORTATION CONTROL INSTRUMENTATION
13600	SOLAR ENERGY SYSTEMS
13700	WIND ENERGY SYSTEMS
13800	BUILDING AUTOMATION SYSTEMS
13900	FIRE SUPPRESSION AND SUPERVISORY SYSTEMS

DIVISION 14—CONVEYING SYSTEMS

14100	DUMBWAITERS
14200	ELEVATORS
14300	MOVING STAIRS AND WALKS
14400	LIFTS
14500	MATERIAL HANDLING SYSTEMS
14600	HOISTS AND CRANES
14700	TURNTABLES
14800	SCAFFOLDING
14900	TRANSPORTATION SYSTEMS

DIVISION 15—MECHANICAL

15050	BASIC MECHANICAL MATERIALS AND METHODS
15250	MECHANICAL INSULATION
15300	FIRE PROTECTION
15400	PLUMBING
15500	HEATING, VENTILATING, AND AIR CONDITIONING (HVAC)
15550	HEAT GENERATION
15650	REFRIGERATION
15750	HEAT TRANSFER
15850	AIR HANDLING
15880	AIR DISTRIBUTION
15950	CONTROLS
15990	TESTING, ADJUSTING, AND BALANCING

DIVISION 16—ELECTRICAL

16050	BASIC ELECTRICAL MATERIALS AND METHODS
16200	POWER GENERATION
16300	HIGH VOLTAGE DISTRIBUTION (Above 600-Volt)
16400	SERVICE AND DISTRIBUTION (600-Volt and Below)
16500	LIGHTING
16600	SPECIAL SYSTEMS
16700	COMMUNICATIONS
16850	ELECTRIC RESISTANCE HEATING
16900	CONTROLS
16950	TESTING

FIGURE 4.3 (*Continued*)

> PART I—GENERAL
>
> Scope; Related Work; Submittals; Inspection Requirements; Testing; Certificates; etc.
>
> PART II—PRODUCTS
>
> Technical specifications for all materials, equipment, fabricated items, etc.; In *no case* is it proper to show any installation requirements in this part, or specify quality of workmanship in this part.
>
> PART III—EXECUTION
>
> Qualitative standards relating to workmanship, etc.; covers installation, erection, construction, etc.; In *no case* is it proper to cover any product, materials, equipment, or fabricated item requirements in this part.

FIGURE 4.4 Three-part section format.

yet see a universal system. Nevertheless, the CSI format is here to stay, and contractors should become familiar with it. Memorize all 16 Division titles, as they never change, even from one job to the next.

CSI Division/Section Concept

Whenever the term *Division/Section Concept* is heard with regard to the CSI format, it is an expression of the relationship of the fixed-title divisions to the subclassifications under each division called *Sections*. Although division titles *never* change from job to job (although some divisions may be omitted from a project if they are not applicable), the titles of the sections that are grouped under them are adapted to the specific needs of each individual project. In Figure 4.3, all 16 fixed-division titles may be used under each appropriate Division. In addition to the short list illustrated, the CSI also publishes a document that provides a ready index of section titles to fit the CSI 16-Division Format.

CSI Three-Part Section Format

One of the most valuable contributions of the CSI to the work of the contractor and the architect-engineer or public agency is the development of the *three-part section format* (Figure 4.4). Under this arrangement, each section is divided into three parts, each containing one type of information only. With this system, fewer items are overlooked simply because the specifications for a particular product were sandwiched between some unlikely paragraphs dealing with the installation of some totally unrelated item, which just happened to be mentioned at that point because some architect or engineer happened to think of it while he or she was writing that portion of the section.

In the three-part section format, all technical sections of the specification are divided into three distinct parts, always in the same order: (1) General; (2) Products; and (3) Execution. If followed faithfully, as most users of the system do, this makes the reading of the specifications a simple, orderly process and eliminates many an error due to oversight.

STATE HIGHWAY DEPARTMENT FORMATS

Long before the coming of the CSI format, the various state highway departments established formats of their own in response to the needs of the type of construction in which they were engaged. Most states have settled on a uniform format based upon the AASHTO (American Association of State Highway and Transportation Officials model).

The basic similarity between all state highway specifications is the fact that they all use a published, bound book of *standard specifications.* This book covers in detail all general contract conditions as well as the technical specifications for all types of construction that could be reasonably anticipated in any highway department project. The subject matter covered is not as narrow as one might at first expect, and to add complications to the specification, it frequently covers several alternative methods of completing the work.

To adapt these standard specifications to a specific project requires an additional document, for the standard specifications themselves cover far too broad a subject area. Furthermore, they do not indicate whether a specific method should be used on a particular project. This adaptation is accomplished by the preparation of a small additional specification called the "Special Provisions" or the "Supplemental Specifications." This document clearly defines the changes to, additions, or deletions from the standard specifications that might be necessary to adapt them to the specific project being constructed. For the sake of uniformity, the Special Provisions follow a standard format adopted by each agency involved, so that all users throughout that state will produce a document that is in the same format,

CHAPTER 4 SPECIFICATIONS AND DRAWINGS

and all contractors will have prior knowledge of the basic requirements and conditions for highway construction in their state.

In the previous cases referred to, when specifications were mentioned, the term was interpreted to include all documents, general conditions, and technical provisions. For a state highway project, this definition must be revised. Here, the specifications are the Standard Specifications, and the document issued for the specific project contains only supplementary material. The usual title for the document containing the front-end documents plus the special technical conditions is *Special Provisions*.

AASHTO Standard Format for Highway Construction Specifications

The majority of the state highway departments of the United States closely follow the standards of the American Association of State Highway and Transportation Officials (AASHTO). An exception to this is the State of California, whose standards were established at an earlier date and retained even after the AASHTO standards came into being.

Rather than referring to the basic or general standards of the AASHTO, the example used here as an AASHTO type specification will be the Road and Bridge Standards of the Virginia Department of Highways and Transportation. Just as in all other states, the Virginia standard specifications cover all potential types of highway and bridge construction that may be encountered, along with all of the possible acceptable alternatives. These must be supplemented by a book of Special Provisions to adapt the Standard Specifications to a particular project. In addition, as a means of keeping the Standard Specifications up to date, the State of Virginia also issues Supplemental Specifications, which are modification sheets needed to update the Standard Specifications. These are normally bound into the set of Standard Provisions.

As with other state highway specifications, the arrangement of the subject matter in the technical portions of the Special Provisions closely parallels that of the contents of the book of Standard Specifications, wherever possible. The Virginia Road and Bridge Specifications are initially divided into seven main divisions.

    I   General Provisions (Sections 101–110; contractual relationships).

STATE HIGHWAY DEPARTMENT FORMATS

    II   Materials (Sections 200–258; for the work of Division II through VII).

    III   Roadway Construction (Sections 301–321; execution only).

    IV   Bridges and Structures (Sections 401–426; execution only).

    V   Incidental Construction (Sections 501–519; execution only).

    VI   Roadside Development (Sections 601–610; execution only).

    VII   Signing (Sections 702 and 703; execution only).

In similar fashion, the Florida Department of Transportation Standard Specifications for Road and Bridge Construction are divided into four main divisions.

    I   General Requirements and Covenants (Sections 1–9; contractual relationships).

    II   Construction Details (Sections 100–715; execution only).

    III   Materials (Sections 901–995; for the work of Division III).

    IV   Contract Forms (printed under separate cover).

As in both the California and the Virginia Standard Specifications, the Florida Standard Specifications divisions are further broken down into sections related to the majority of construction materials and methods expected to be encountered in normal road and bridge construction. Anything needed for a project that is not contained in the standard specifications is specified in the Special Provisions.

What all state highway specifications have in common with the CSI Format (long before the formation of the CSI) is the strict separation of materials and execution in their specifications. Thus, in all execution portions of a state highway specification, the materials are covered by reference to the detailed specifications in the division provided for that purpose.

The numbering of the special provisions technical sections in the Virginia highway specifications uses a fixed number system that retains the numbers used in the original book of Standard Specifications, unlike the California example next described, in which a fixed numbering system is used that is wholly independent from the numbering system in its standard specifications.

CHAPTER 4 SPECIFICATIONS AND DRAWINGS

California Department of Transportation Format

The California Department of Transportation or "Caltrans" standard specifications were designed to meet the specific needs of the construction industry in California for street, highway, and bridge construction projects and as such were adopted by many cities and counties throughout the state who administer similar projects at local levels.

The Standard Specifications must be supplemented by a book of Special Provisions as required for all highway projects. The format used in these special provisions has been standardized, and a uniform format with fixed-number sections and standard section titles has been adopted in California.

The arrangement of a typical California Department of Transportation Special Provisions document is shown in Figure 4.5. The list of subjects

---

NOTICE INVITING BIDS

SPECIAL PROVISIONS

Section 1 —Specifications and Plans (definitions)
Section 2 —Proposal Requirements and Conditions
Section 3 —Award and Execution of Contract
Section 4 —Beginning of Work; Time of Completion; and Liquidated Damages
Section 5 —General (includes updates to Standard Specifications)
Section 6 —(Content varies)
Section 7 —Legal Relations and Responsibility
Section 8 —Miscellaneous (technical data)
Section 9 —Description of the Work (definition of scope)
Section 10—Construction Details (technical provisions)

PROPOSAL AND CONTRACT

Proposal and Bid Sheets
Bid Bond
List of Subcontractors
Agreement
Performance Bond
Payment Bond

FIGURE 4.5 California format for special provisions.

covered, as well as the standardized title of each section, remains unchanged from one project to the next.

NONSTANDARD CONSTRUCTION SPECIFICATION FORMATS IN USE

There are several types of approaches to the problem of separating a project specification into seemingly logical units of construction. The two most common concepts are

1. Separation into trade-group and material classifications, as in the CSI format.
2. Separation by construction features, wherein each significant feature of a project is described completely within a single section, including all the materials and methods involved to complete the specified structure or feature.

Depending on the nature of a specific project, each concept may have something good to be said of it, and each could serve to special advantage if used properly. If the two systems were ever mixed in the same specifications, however, the job of field administration would become chaotic! The problems resulting from such an unwise choice include the inability to reasonably control payments to the contractor for the various portions of the work, difficulty in defining interfaces in construction, and duplication of specification provisions in different parts of the work, often with varying and conflicting requirements. A contractor who is awarded a project like this should be prepared for a lion-sized job of maintaining cost control, especially if the job turns out to be a unit price project.

Figure 4.6 provides a graphic example of the complicated relationships involved; it takes the various features of a recreational park project as an example and superimposes them in matrix form with the various trade-group and material classifications that apply to the same work.

It becomes obvious from the chart that any project including all these various features would be most benefited by the CSI Format, which is based upon the separation of the specification technical provisions into sections corresponding to the *vertical* columns in the chart in Figure 4.6. Thus, there would be only one concrete section, one metals section, one electrical section, and so forth no matter how many separate structures or features were included in the project.

Item	Unit of Construction	CSI Division Categories														
		2	3	4	5	6	7	8	9	10	11	12	13	14	15	16
A	Site Grading	x														
B	Site Utility Lines	x	x		x										x	
C	Storm Drain System	x	x		x											
D	Valve Chamber	x	x	x	x		x		x		x				x	x
E	Streets & Parking	x	x		x				x						x	x
F	Snack Bar Building	x	x	x	x	x	x	x	x	x	x				x	x
G	Rest Room Building	x	x	x	x	x	x	x	x	x	x				x	x
H	Administration Bldg.	x	x	x	x	x	x	x	x	x	x				x	x
I	Landscaping & Irrig.	x	x												x	

FIGURE 4.6 Comparison of trade group vs. project feature formatting.

PROJECT SPECIFICATIONS VS. SPECIAL PROVISIONS CONCEPT

This subject has been slightly addressed in earlier paragraphs, but it is of great importance for the contractor to know the relative importance of the documents that must be used in the field and to understand the very important difference between these two concepts.

Project Specifications (CSI Project Manual)

The *project specifications* (or CSI Project Manual) concept is based on the issuance of a single, all-inclusive printed volume containing all of the contract provisions that apply to the job (Figure 4.7), although *references* to outside sources are permissible. The effect of such outside references in this type of specification, however, is to bind the contractor only to the extent of the specific reference specification named. Thus, if a project specification states that "portland cement shall be Type II cement as specified in Section 90 of the Caltrans Standard Specifications," it limits the control of the cited reference to the *portland cement specifications* only and does not bind the contractor to any other provisions of that same section, such as the grading of aggregates for concrete, and so on.

PROJECT SPECIFICATIONS VS. SPECIAL PROVISIONS CONCEPT

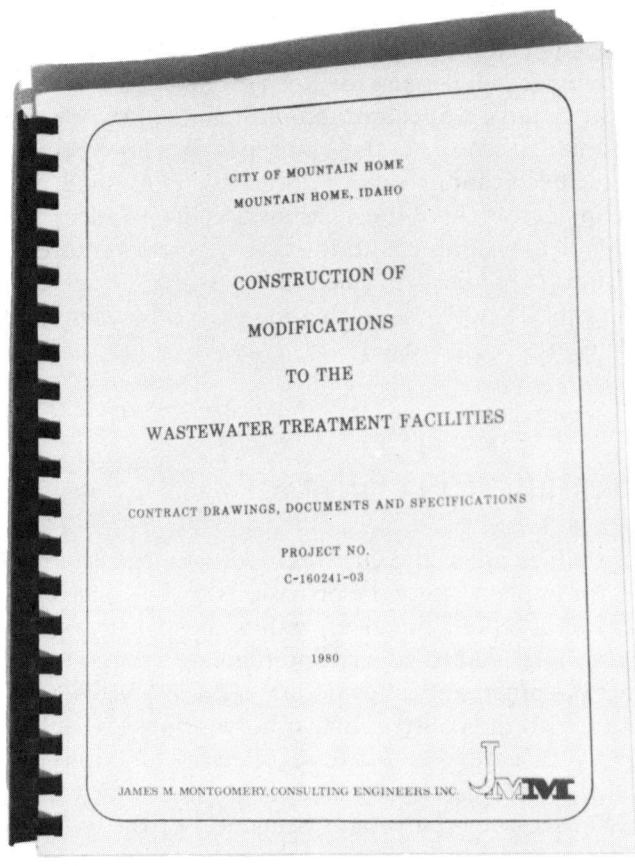

FIGURE 4.7 Comparison of job specifications at left, with standard specifications plus special provisions at right, for similar projects.

Special Provisions or Supplemental Specifications

The *special provisions* or supplemental specifications concept is based upon the idea that a previously published book of Standard Specifications is the actual detailed specification for all applicable work on the project, and that the Special Provisions or Supplemental Specifications are merely a supplemental document to provide for those items on a particular project that the design engineer wanted to change from the provisions of the Standard Specifications, or to make a specific selection of options provided in the Standard Specifications. For example, although it has already been established that the entire concrete section of the Standard Specifications will control the project (insofar as it is applicable, of course), the reference in the Special Provisions or Supplemental Specifications that "portland cement shall be Type II cement as specified in Section 90 of the Caltrans Standard Specifications" merely controls the choice of option as to the *type of cement* required for the work. All the provisions of the rest of "Section 90" still apply to the concrete work. An exact specification phrase in the *project specifications* can have a vastly different meaning when used in a *special provisions* or *supplemental specifications* document.

GENERAL CONDITIONS OF THE CONSTRUCTION CONTRACT

The *General Conditions,* sometimes called the *General Provisions,* specify the manner and the procedures for implementing the provisions of the construction contract according to the accepted practices within the construction industry. These conditions are intended to govern and regulate the requirements of the formal contract or agreement. They do not serve as a waiver of any legal rights that either party to the contract may otherwise possess. The General Conditions of the Contract are intended to regulate the functions of either party only to the extent that each party's activities may affect the contractual rights of the other party or the proper execution of the work.

Although the general content of the general conditions of most public works construction contracts varies somewhat from one set to another, depending upon the requirements of the agency that originated the document, the following list taken from the General Conditions of the "Contract Documents for Construction of Federally Assisted Water and Sewer Projects," Document No. 11 may well be considered as somewhat typical.

1 Definitions (terms used in the contract documents)
2 Additional Instructions and Detail Drawings (legal status of such)
3 Schedules, Reports, and Records (submittals by the contractor)
4 Drawings and Specifications (intent; order of precedence; conflicts)
5 Shop Drawings (contractor's obligations; procedures for handling)
6 Materials, Services, and Facilities (responsibility for)
7 Inspection and Testing

GENERAL CONDITIONS OF THE CONSTRUCTION CONTRACT

8 Substitutions (handling of "or-equal" products)
9 Patents (Payment of royalties; hold harmless provisions)
10 Surveys, Permits, Regulations (who furnishes and who pays)
11 Protection of Work, Property, Persons (responsibility for)
12 Supervision by the Contractor (requirements for)
13 Changes in the Work (change orders)
14 Changes in Contract Price (as the result of change orders)
15 Time for Completion and Liquidated Damages
16 Correction of Work (correction of rejected work)
17 Subsurface Conditions (discovery of unforeseen conditions)
18 Suspension of Work, Termination, and Delay
19 Payments to Contractor (procedures)
20 Acceptance of Final Payment as Release
21 Insurance (P.L. & P.D.; worker's comp.; builder's risk; fire)
22 Contract security (performance and payment bonds)
23 Assignments (no assignment of contract)
24 Indemnification (hold harmless)
25 Separate Contracts (owner's right to have)
26 Subcontracting (must be modified in some states)
27 Engineer's Authority
28 Land and Rights-of-Way (obligations to provide)
29 Guaranty (work and materials)
30 Arbitration (agreement to arbitrate future disputes, varies in some states)
31 Taxes (who pays)

Other standard sets of General Conditions have been developed by various segments of the construction industry. The American Institute of Architects (AIA); the Engineer's Joint Contract Documents Committee (EJCDC); and various branches of the federal, state, and municipal governments have all developed and published standardized sets of General Conditions.

Although most of these documents bear a superficial similarity to one another, certain federally produced documents depart significantly from the normal industry standards.

In concept, standard forms of General Conditions have many advantages. Not only are they generally the result of collaboration with industry leaders and thereby represent a fair and equitable method of handling a construction contract, but they have normally had the advantage of being thoroughly critiqued by members of the architect-engineering profession, the legal profession, and contractor organizations such as the Associated General Contractors (AGC). Furthermore, many have been tested in court and can be relied upon to provide similar protection to all who use them. Another and frequently overlooked advantage is that many such documents have evolved into a form that has withstood the test of time and experience and have become familiar to the contractors who use them. After repeated use, a contractor will clearly understand all of a document's terms, meanings, and implications. If documents are good, this is often reflected in the stability of bid prices, as repeated prior use enables a contractor to fairly assess the full effect of the contract provisions on the conduct of the work.

It must be remembered, however, that because a document *is* so "general" in nature, certain types of specific data cannot be included, otherwise it would not be applicable for all projects. For example, a provision in the General Conditions for "liquidated damages" or "bonds" will normally only cover the limitations, procedures, or other unchanging elements, but *not* the actual monetary amount. The specific dollar amounts must be referred to in the Supplementary General Conditions, a document that will be discussed later.

Obviously, there is no shortage of "standards" and the chance of encountering any particular standard on any given project may be influenced by many things, including the specific public agency requirements, source of construction funds, type of work to be constructed, whether the project is being built with local or federal funds, and last but not least, whether a particular set of General Conditions contains provisions that the public agency and its architect-engineer are willing to accept.

Given the multitude of standard General Conditions available, there would seem to be little valid reason to generate a new one, yet many architect-engineers and public agencies are doing just that when they create their own General Conditions to fit a particular job. Each time that a new set is written, there is always the danger that its provisions may contain subtle wording that may contain new, unforeseen risks to the unwary contractor.

General Conditions Provisions of Standard Specifications

On some projects for which the basic contract is written around a set of Standard Specifications, the General Conditions of the Contract are usually contained in the early chapters of the Standard Specifications book. For example, a project for the construction of a bridge under the Virginia Road and Bridge Specifications has its General Conditions specified in Division I,

"General Provisions," which includes Sections 101 through 110 of that document.

Similarly, the American Public Works Association in cooperation with the Associated General Contractors published a book of standard specifications entitled "Standard Specifications for Public Works Construction," usually simply referred to as the *Green Book*. Here, as in the standard state highway specifications, the General Conditions are actually contained in Part 1, "General Provisions," Sections 1 through 9 of the standard specifications book. The principal difference between the two books lies in two areas: (1) The Green Book contract provisions are designed to accommodate a broader type of construction, whereas the state highway and bridge specifications are primarily designed for highways, bridges, and drainage facilities; and (2) the state highway standard specifications are designed for use only in one state and therefore contain all of the special legal requirements that must be a part of public contracts in that state. The Green Book was originally written for use in any state, however, the recent inclusion of California code requirements seriously limits its use in other states.

SUPPLEMENTARY GENERAL CONDITIONS

As the name implies, the Supplementary General Conditions are simply an extension of the General Conditions of the Contract. It is in this document that the special legal requirements of the contract are expanded to include provisions that apply solely to the project at hand. In some cases, the titles of Articles within the Supplementary General Conditions will duplicate titles already mentioned in the General Conditions. This is neither repetitious nor necessarily a superseding provision. In most cases, both such paragraphs still apply. However, the provisions in the General Conditions may contain only general or procedural responsibility clauses, whereas the same subject if covered in the Supplementary General Conditions (or *Special Conditions* as it is named in some documents) will add specific requirements that apply only to the subject job, such as the *amounts* of liquidated damages, bonds, or insurance required.

As previously mentioned, the Supplementary General Conditions portion of the Contract Documents may appear under any one of several different names, but it is still the same document. Often, under older formats, this portion of the specifications is referred to as *Special Conditions* or on some of the newer formats as Supplementary General *Provisions* instead of Conditions.

In addition to the items that are expansions of the Articles already specified in the General Conditions, there are numerous other subjects that may be encountered in the Supplementary General Conditions. A sample listing of the contents of one such public agency document follows:

1 Scope [of the entire project]
2 Supplementary Definitions [not already covered in the General Conditions]
3 Legal Address of the architect-engineer and Owner [needed for service of legal notices]
4 Amounts of Bonds [actual dollar or percentage values]
5 Amount of Liquidated Damages [actual dollar value]
6 Permits and Inspection Costs [who pays what?]
7 Contract Drawings [complete listing, by number and title, of all drawings that are made a part of the contract]
8 Applicable Laws and Regulations [specific requirements for the job]
9 Insurance [amount of coverage; additional insurance not specified in the General Conditions and amount of its coverage]

TECHNICAL PROVISIONS OF THE SPECIFICATIONS

It should be noted that two of the three parts of the typical specifications document have "general" clauses in them. They are Part II, "Conditions of the Contract" and Part III, "Technical Provisions." Some distinction should be noted between these two portions of the specifications to avoid confusion.

First, the general provisions of the *Conditions of the Contract* relate to the contractual relationships and legal obligations of the parties to the contract. The general provisions of the *technical* portion of the specifications should relate to those requirements of a technical nature that apply generally to the work of the entire project, rather than to the work of one trade. Thus, the General Conditions refer to legal contractual issues, whereas the General Requirements refer to construction details, project features, procedural requirements, submittals, scheduling constraints, and similar project-related functions of a general nature.

Under the CSI Format, Division 1 of the 16-Division format was reserved for this purpose. The types of subjects generally included, where applicable, are the following:

TECHNICAL PROVISIONS OF THE SPECIFICATIONS

GENERAL REQUIREMENTS—DIVISION 1

1. Summary of Work
    - Work by Others
    - Items Provided by Owner
    - Work Included in this Contract
    - Work to be Performed Later

2. Alternatives

3. Measurement and Payment

4. Project meeting
    - Preconstruction conferences
    - Progress meetings
    - Job Site administration

5. Submittals
    - Construction schedules
    - Network analyses
    - Progress reports
    - Survey data
    - Shop drawings, product data, & samples
    - Operation and maintenance data
    - Layout data
    - Schedule of values
    - Construction photographs

6. Quality control
    - Testing laboratory services
    - Inspection services

7. Temporary facilities and controls
    - Temporary utilities
    - Construction elevators and hoists
    - Guards and barricades
    - Shoring, falsework, bracing, scaffolding, and staging
    - Access roads
    - Control of dust, noise, water, vapors, and pollutants
    - Traffic control, parking, storage of materials
    - Temporary field offices

8. Products
    - Quality
    - Transportation and handling
    - Storage and protection

9 Project closeout
      Cleaning up
      Project record documents (as-built drawings)
      Touch-up and repair
      Operational testing and validation
      Maintenance and guaranty
      Bonding requirements during guarantee period

These and similar provisions are generally representative of the types of subject matter that are expected to be contained in Division 1 of the technical portion of the specifications when the CSI Format is being used. A comparison of these subjects with the subject area covered under the "General Conditions of the Contract," Part II should give some idea as to the usual arrangement of the various subject matter.

Provisions for Temporary Facilities

Most public works specifications will include requirements for certain temporary facilities and services that must be provided by the contractor by a specific date.

It should be emphasized that when temporary facilities and services are specified, it becomes the contractor's basic responsibility to provide all plant and equipment that is necessary to provide adequately for the performance of the contract, within the time specified. All such plant and equipment must be kept in satisfactory operating condition and must be capable of safely and efficiently performing the required work. Sometimes, such items are specified to be subject to the inspection and approval of the architect-engineer or its designated resident project representative.

A number of specific issues may be encountered in the specifications to facilitate these requirements. The following is a list of items that may be expected to be encountered in a public works contract, when applicable to the type of project being constructed:

Temporary electric services
(Includes installation of all construction lighting, wiring, and circuit separation and the payment of usage charges by the utility companies)

Fire protection
(Includes connections to water supply system)

Temporary utility services
(Includes water supply development and connection and later removal of

such connections; power supply and connections; free local telephone service for both the contractor's and the public agency's or architect-engineer's field offices; type of telephone service required and cost of toll calls; and time of installation of such facilities)

Sanitation
(Includes toilet facilities and disposal of waste materials)

Site access and storage provisions
(Includes highway limitations; marine transportation provisions, including pier and landing facilities and small boat launching facilities; and contractor's work and storage area limitations)

Environmental controls
(Includes explosives and blasting limitations, dust abatement, chemical use and disposal, and misplaced or discharged materials into navigable waterways)

Cultural resources
(Includes historical, architectural, archaeological, or other cultural resources endangered by the project; the owner's right to stop the work in case of a cultural resource "find.") [Such a work stoppage may provide a good chance of recovery by the contractor as a compensable delay, however; provided that the terms of the contract allow it.]

Field office facilities
(Includes type of structure or facility required; office equipment and furnishings to be included; date of installation and completion of field office facility; and cleanup services required.)

As a part of the specifications requirements for temporary facilities to be provided by the contractor, some of these items may actually be shown as a separate pay line item on a unit price bid, or as part of a pay line item entitled *mobilization*. In this way, failure by the contractor to provide such temporary facilities on a timely basis can delay the contractor's first progress payment.

ADDENDA TO THE SPECIFICATIONS

Addenda to the specifications, or in its singular form, an addendum, to a particular set of specifications is a document setting forth the changes, modifications, corrections, or additions to the contract documents that have been issued after the project has been advertised for bids, *but before the time of*

THE METROPOLITAN WATER DISTRICT
OF SOUTHERN CALIFORNIA

Supplement No. 1 to Notice Inviting Bids
under Specifications No. 1121

To all prospective bidders under Specifications No. 1121 for constructing the Victoria Street–223rd Street Cross Feeder Relocation, for which bids are to be received by the Metropolitan Water District of Southern California at its office at 1111 Sunset Boulevard, Los Angeles, California (P.O. Box 54153, Los Angeles, California 90054) until 10:00 a.m., May 22, 1985.

I. Metropolitan Water District Standard Specification C11, Polyethylene Coatings, attached hereto, shall be included in Appendix A of the above-described specifications.

II. Without issuing a revised drawing, the following changes are hereby made on Drawing No. B-69201, Sheet No. T-11, forming a part of these specifications.

A. In Details 1, 2, 3, 4, 5, and 9, requirements for Carnegie M-3818 spigot rings shall be changed to Carnegie M-3516 spigot rings.

B. In Detail 3, the rubber gasket diameter shall be changed from 13/16 inch to 21/32 inch.

III. Specifications No. 1121 as originally issued shall be used in submitting bids and a copy of this Supplement No. 1 to Notice Inviting Bids shall accompany the bid.

THE METROPOLITAN WATER DISTRICT
OF SOUTHERN CALIFORNIA

By
Carl Boronkay
General Manager

May 10, 1985

FIGURE 4.8 Specifications addendum.

opening bids. It is hoped that such a document is issued sufficiently in advance of the bid opening date to allow bidders to make the necessary changes in their bids. In some areas, the law requires that an addendum be issued a set number of days before the bid opening date.

The addenda may be referred to as *addendum to the specifications* or as an *addendum to the Notice of Inviting Bids.* Each has the same legal effect. Many public agencies seem to prefer the latter (see Figure 4.8). The addenda must be delivered to each party who has obtained a set of bid documents.

One of the first things that a contractor should do on receipt of an addendum is to carefully check the specifications and drawings and mark all corrections, changes, modifications, or additions to the original documents. The next step is to cross check all documents to see if any of the data that was changed involved an interface with other sections of the specifications or any other drawings. Although normal errors and omissions to the original design drawings and specifications may provide the contractor with a sound basis for a claim of extra cost, *apparent conflicts resulting from the issuance of addenda do not.* This is simply because the addenda take precedence over all other documents, and the contractor is traditionally expected to adapt the changes to the work. The contractor should review the entire document for any such impacts and mark its own set of contract documents to reflect all changes by addenda.

It should be remembered that addenda can only be issued during the bidding period on public works contracts. Any changes that are made by the architect-engineer after a bid opening must be deferred until after the signing of an agreement and then must be executed as a change order (see Figure 4.9). This will also give the contractor a chance to negotiate the cost in both time and money.

STANDARD SPECIFICATIONS

By definition, a set of *standard specifications* is a prepublished set of specifications, usually in book form, that normally contains both the conditions of the contract (boilerplate) and technical provisions for all types of construction and materials that the originating agency expects to cover during any of its projects. When adopted by a public agency or an architect-engineer working on a project for a public agency, the total content of this becomes a part of the Contract Documents, subject only to changes set forth in a separate project-specific document called Special Provisions or *Supplemental*

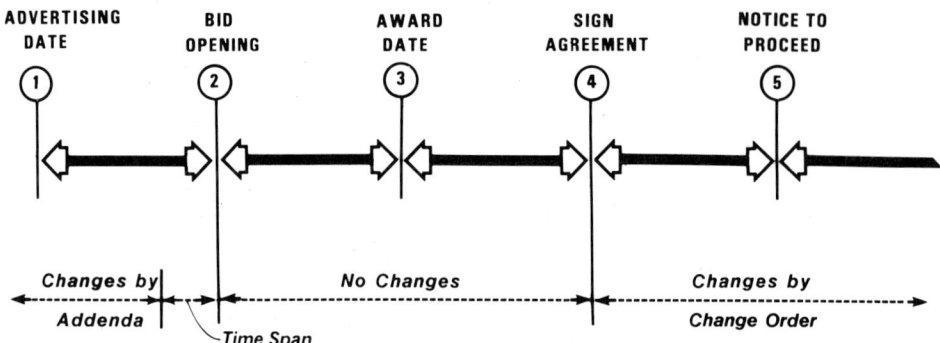

FIGURE 4.9 Precontract change procedure.

Specifications that adapts the somewhat general nature of the standard specifications to the specific needs of a particular project. In this manner, wherever alternatives are offered in the Standard Specifications, the Special Provisions or Supplemental Specifications serve to indicate which of the available choices apply to this specific project. Under this type of contract, anything not modified by the terms of the Special Provisions or Supplemental Specifications is required to comply with *all* applicable provisions of the Standard Specifications.

In direct contrast to this approach, under the Project Specifications (Project Manual) format, if a project does not cite the Standard Specifications as the principal contract document, but merely references certain sections of it, then nothing in the Standard Specifications will apply to the subject project except those items specifically referenced in the Project Specifications or Project Manual. The contractor must be very careful with regard to citations from Standard Specifications to assure proper understanding of the contractual obligations.

Particular attention should be paid to the exact numerical designation of a citation. For example, if the citation read

. . . per Section 223.02(d) of the Virginia Road and Bridge Specifications . . .

Such a specific reference would preclude the use of anything that is not covered in that particular subsection (d). In this case, it would mean that the concrete must be cured solely by the use of the "liquid membrane seal" method, which is specified in detail in the Standard Specifications.

On the other hand, if the citation read

. . . per Section 223 [or 223.02] of the Virginia Road and Bridge Specifications . . .

then the contractor would be free to select any one of four specified acceptable methods of curing the concrete; it would be the contractor's option.
Under the California Department of Transportation specifications, the coverage of each section is broader; Section 90 of that document includes not only portland cement, but also aggregates, curing materials, admixtures, and similar items. The effect of making a reference as general as the following

. . . per Section 90 of the Caltrans Standard Specifications . . .

in a contract where the book of Standard Specifications is the principal document would be to require the contractor's compliance with *all* of the provisions of that Section, including a choice of each alternative provided. If the engineer wanted to limit the use of curing methods, for example, the Special Provisions or Supplemental Specifications would have to include a qualifying statement to that effect.
The use of a set of Standard Specifications without an accompanying set of Special Provisions or Supplemental Specifications is like asking to read a copy of a book and being handed a dictionary instead and being told "All the words are here; just read the ones that apply."
Normally, standard specifications will only be used as a primary specification document on engineering projects, as the provisions seldom cover the subject areas needed for building construction. However, cross references to limited portions of a standard specification are frequently made in all types of construction.

SPECIAL MATERIAL AND PRODUCT STANDARDS

If every time that an item was specified for a project, the specifications would have to contain all of the provisions that were necessary to assure that the product met all physical, chemical, geometrical, or performance standards required, every set of project specifications would have to be from 10 to 100 times bulkier than it is now. Some would look like a set of encyclopedias. Worse, every architect-engineer without the benefit of coordination would have enough subtle differences in its respective descriptions of the same product that the manufacturers would be solely in the business of producing custom-made materials for every different project. Even if this could be successfully accomplished, the construction costs would skyrocket to astronomical levels.

As a means of providing the uniformity necessary, various nonprofit trade associations as well as governmental agencies and manufacturers have established voluntary standards that are actually standard specifications for separate, individual products. Thus, by referring to the published data for each of these products, a public agency or architect-engineer can design a project subject to the specified product limitations, with full assurance that such products are not only available on the market, but are carefully regulated by each manufacturer to assure compliance with the previously established standards.

The agencies that issue such standards are sometimes governmental, sometimes industry trade associations, and sometimes independent standards associations whose only function is the preparation of such industry standards—with the voluntary cooperation of industry, of course. In each case, the standards established represent the coordinated efforts of the manufacturer, the architect-engineers, the academic community, and other parties, as applicable.

Such standards become a part of a construction contract only if specifically called out in the specifications and drawings and then only to the extent of the specific reference. If a specification calls for a particular product by its ASTM designation, but includes something that was not a part of that ASTM standard, the product must conform to the cited ASTM standard *subject to the modifying provisions*. It would thus actually be a "specially modified" product, requiring the manufacturer to produce a custom item with appropriate increase in cost and delay in the delivery schedule.

Such standards may be loosely divided into two basic classifications: (1) Government Standards and (2) Nongovernment Standards. In the first category, the following are most commonly used.

Government Standards

Federal Specifications

Federal specifications describe essential and technical requirements for items, materials, or services that are normally bought by the federal government. They are also extensively referred to in specifications for nonfederal projects when commercial standards are not available for a particular item. They are generally characterized by an alpha-numeric system; for example, "SS-C-1960/3B" for Portland Cement, as illustrated in Figure 4.10.

178	NUMERIC LIST OF FEDERAL SPECIFICATIONS & COMMERCIAL ITEM DESCRIPTIONS					
DOCUMENT NUMBER	QPL	TITLE	FSC	PREP	DATE	PRICE
RR-W-670D		Wringer, Mop	7920	FSS	17 Oct 72	
RR-W-1101A		Waste Receptacle, Swinging Doors	7240	FSS	31 Oct 72	
RR-W-001588		Waste Receptacle, Wall Or Post Mounted	7240	FSS	20 Mar 70	
RR-W-1817A		Warning Device, Highway, Triangular, Reflective	9905	FSS	30 Jul 75	
SS-A-281B(1)		Aggregate; (For) Portland-cement-concrete	5610	ME	25 Jan 57	
SS-A-666D NOTICE 1		Asphalt, Petroleum (Built-up Roofing, Waterproofing, And Dampproofing)	5610	FSS	7 May 68 24 Jul 68	
SS-A-671C		Asphalt, Liquid; Slow-curing, Medium Curing, And Rapid Curing	5610	FSS	8 Sep 66	
SS-A-694D		Asphalt Roof Coating (Brushing And Spraying Consistency)	5610	FSS	9 Mar 73	
SS-A-701B		Asphalt, Petroleum (Primer, Roofing, And Weatherproofing)	5610	FSS	15 Jul 74	
SS-A-706D		Asphalt, Petroleum' road And Pavement Construction (Asphalt Cement)	5610	YD	25 Sep 78	
SS-B-656B		Brick, Building, Common (Clay Or Shale)	5620	ME	18 Feb 66	
SS-B-663B		Brick, Building, Concrete	5620	FSS	9 Jun 65	
SS-B-668B		Brick, Facing, Clay, Or Shale	5620	FSS	20 May 74	
SS-B-671C		Brick, Paving	5620	FSS	2 Jun 65	
SS-B-681B		Brick, Building, Sand-lime	5620	FSS	22 Dec 65	
SS-B-755A(1)		Building Board, Asbestos Cement' flat And Corrugated	5640	FSS	21 May 68	
SS-C-153C		Cement, Bituminous, Plastic	5610	FSS	13 Dec 74	
SS-C-160A(2)		Cements, Insulation Thermal	5640	FSS	11 Jan 79	
SS-C-161A		Cement' keene's	5610	FSS	23 Jan 74	
SS-C-255		Chalk, Carpenters' And Railroad	7510	FSS	30 Aug 49	
SS-C-00255A		Chalk, Carpenters' And Railroad	7510	FSS	10 Jan 69	
SS-C-266F		Chalk, Marking, White And Colored	7510	FSS	20 Aug 73	
SS-C-450A		Cloth, Impregnated (Woven Cotton Cloth, Asphalt Impregnated; Coal Tar Impregnated)	5650	FSS	2 Jul 64	
SS-C-466E INT AMD 2		Cloth, Thread, And Tape; Asbestos	5640	SH	2 Jul 64 22 Sep 76	
SS-C-540B		Coal Tar (Cutback) Roof Coating, Brushing Consistency	5610	FSS	11 Jan 75	
SS-C-621B INT AMD 2		Concrete Masonry Units, Hollow (And Solid, Prefaced And Unglazed)	5620	FSS	19 Jun 68 18 Jun 70	
SS-C-635B		Crayon Assortment, Drawing, Colored	7510	FSS	7 Dec 66	
SS-C-646B(2)		Crayon, Marking, Lumber	7510	FSS	31 Oct 63	
SS-C-661A		Crayon, Marking	7510	FSS	11 Jan 74	
SS-C-1783		Cloth, Asbestos	5640	FSS	29 Mar 73	
SS-C-1960/GEN		Cement And Pozzolan (General Requirements For)	5610	YD	2 Dec 75	
SS-C-1960/1A		Cement, Masonry	5610	YD	24 May 77	
SS-C-1960/2A		Cement, Natural	5610	YD	28 Aug 78	
SS-C-1960/3B		Cement, Portland	5610	YD	28 Aug 78	
SS-C-1960/4B		Cements, Hydraulic, Blended	5610	YD	24 May 77	
SS-C-1960/5A		Pozzolan, For Use In Portland Cement Concrete	5610	YD	28 Aug 78	
SS-F-001032		Floor Covering, Asphaltic Felt (Bituminous Type Surface)	7220	FSS	19 Oct 66	
SS-G-659A		Graphite, Dry (Lubricating)	9620	FSS	1 Mar 67	
SS-J-570B		Joint Compounds And Tape, Wallboard (For Gypsum Wallboard Construction)	5640	FSS	9 Aug 77	
SS-L-30D INT AMD 3		Lath, And Board Products, Gypsum	5640	FSS	27 Feb 74	

FIGURE 4.10 Example of federal specifications index listing.

CHAPTER 4 SPECIFICATIONS AND DRAWINGS

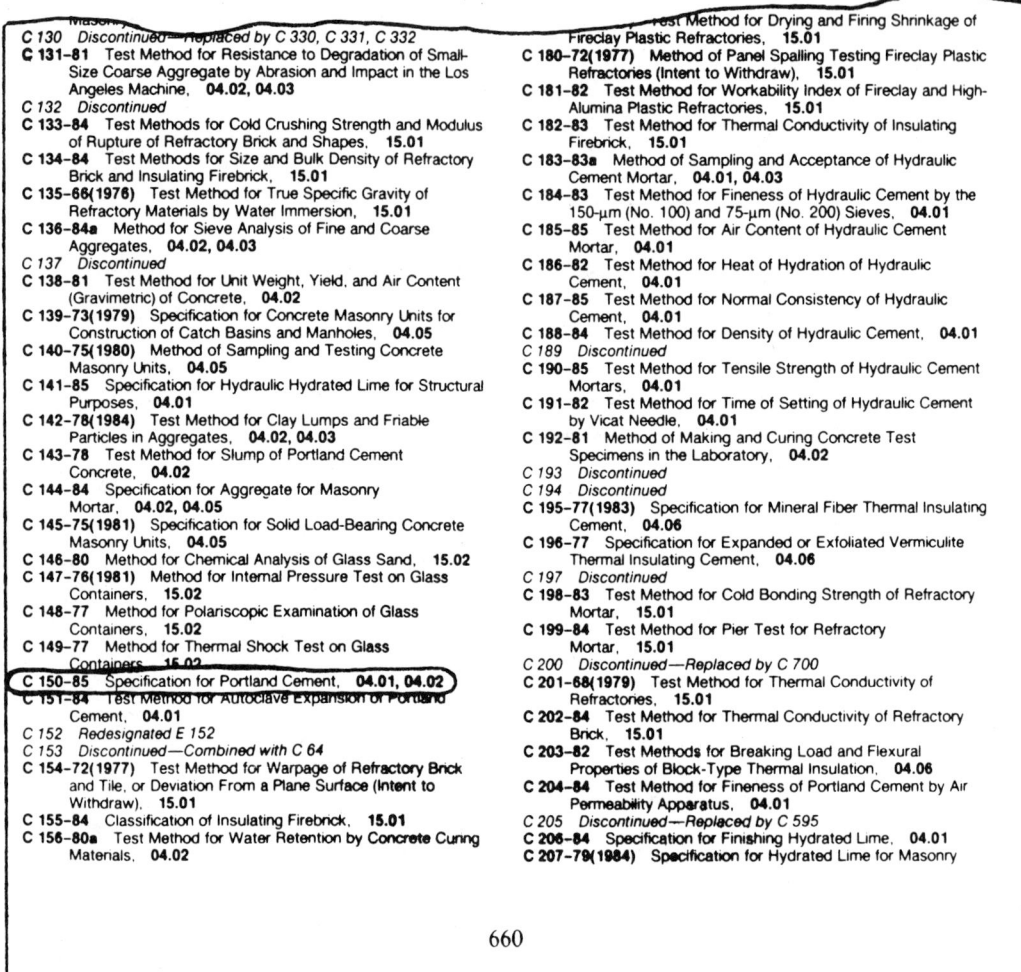

FIGURE 4.11 Example of page from ASTM Index of Standards.

Military Specifications (Department of Defense)

Military specifications, or "Mil-Specs" as they are often called, specify products that are usually unique to the needs of the military; however, in some rare cases, such as certain electrical devices or metal alloys like monel, they may be the only source of an appropriate material specification. They are

SPECIAL MATERIAL AND PRODUCT STANDARDS

generally not used in civil engineering or architectural projects because of the difficulty in obtaining copies of the standards as well as the restricted availability of manufacturers who produce these standards. Mil-Specs are characterized by designations such as "MIL-R-0039016A," the designation for an electrical relay.

UBC Standards

Although technically not a governmental agency, the UBC Standards, whenever referred to, are actually backed up by local city or county ordinances, thus carrying the force of law. All cities and counties who have adopted the Uniform Building Code include UBC Standards as a part of their requirements by virtue of the fact that they are covered by reference in the Uniform Building Code. The majority of UBC Standards are in fact other commercial standards that have been adopted as UBC Standards and renumbered. For example, the UBC designation for Portland cement is "UBC Standard 26-82." This standard is based upon ASTM Designation C 150-81.

Nongovernment Standards

ASTM (American Society for Testing and Materials)

By far the most recognized of all American standards, ASTM covers not only materials specifications, but testing requirements, and in some rare cases performance standards as well. Its listings are characterized by references such as "ASTM Specification C 150-84," with the "84" denoting the year of the particular edition or revision of that particular standard as illustrated in Figure 4.11, which shows a sample page from the ASTM Index of Standards. All ASTM Standards are divided systematically into groupings with separate letter prefixes that enable a user to identify the type of material referred to from the letter designation alone. The following is a complete listing of all current letter prefixes and the material categories they refer to:

- A. Ferrous metals.
- B. Nonferrous metals.
- C. Cementitious, ceramic, concrete, and masonry materials.
- D. Miscellaneous materials.

E. Miscellaneous subjects.
F. Materials for specific applications.
G. Corrosion, deterioration, and degradation of materials.
ES. Emergency standards.

ANSI (American National Standards Institute)

ANSI is a coordinating organization for a federated national standards system. The ANSI federation consists of companies, large and small, and some trade, technical, professional, labor, and consumer organizations. In cooperation with its councils, boards, and committees, ANSI coordinates the efforts of many organizations in the United States that are developing standards.

ANSI does not develop any standards of their own, but it does provide the means for determining the need for them and assures that organizations competent to fill these needs undertake the standards development work. Most of the standards listed by ANSI are developed by trade, technical, professional, consumer, and labor organizations. Each of these standards is then submitted to ANSI for recognition as a national consensus standard.

The System of Standards identification used by ANSI prior to 1979 is no longer in use, and currently, all ANSI Standards are identified by the sponsor's own numbering system, prefixed with the letters "ANSI" followed by the letter or number designations assigned that standard by the sponsoring agency. This is illustrated in Figure 4.12 that displays a page from the ANSI standards index.

AWWA (American Water Works Association)

Not as well known as ASTM or ANSI, except in public works construction, AWWA has been the standard for the water industry. Its standards are broad, and many specify a single fabricated item or an entire project feature, such as the designation "AWWA D100" [the full designation is ANSI/AWWA D100-84 (AWS-D5.2-84)] for the design and construction of welded-steel elevated water tanks, standpipes, and reservoirs. The AWWA publishes its data in several forms, including reference books, handbooks, manuals, standards, periodicals, and pamphlets. Some AWWA Standards have also been adopted by ANSI, as in the previous example of standard D100.

Abrasion Resistance of Concrete or Mortar Surfaces by the Rotating Cutter Method, **ANSI/ASTM C944-80,** ▲

Compressive Strength of Concrete Cylinders Cast in Place in Cylindrical Molds, **ANSI/ASTM C873-80,** ▲

Compressive Strength of Concrete Using Portions of Beams Broken in Flexure, Method of Test for, **ANSI/ASTM C116-68(1980),** ▲

Compressive Strength of Lightweight Insulating Concrete, Test for, **ANSI/ASTM C495-80,** ▲

Consolidation of Concrete, Practice for, **ANSI/ACI 309-72(1978),** $13.95

Epoxy-Resin Systems Used with Concrete, Test for Effective Shrinkage of, **ANSI/ASTM C883-80,** ▲

Evaluation of Strength Test Results of Concrete, Recommended Practice for, **ANSI/ACI 214-77,** $9.00

Foaming Agents for Use in Producing Cellular Concrete Using Preformed Foam, Testing, **ANSI/ASTM C796-80,** ▲

Foaming Agents Used in Making Preformed Foam for Cellular Concrete, Specifications for, **ANSI/ASTM C869-80,** ▲

Freezing and Thawing, Test for Resistance of Concrete to Rapid, **ANSI/ASTM C666-80,** ▲

Making Test Cylinders and Prisms for Determining Strength and Density of Preplaced-Aggregate Concrete, Method for, **ANSI/ASTM C943-80,** ▲

Making, Curing, and Testing Concrete Compression Test Specimens at an Early Age, **ANSI/ASTM C918-80,** ▲

Measuring, Mixing, Transporting, and Placing of Concrete, Practice for, **ANSI/ACI 304-73(1978),** $22.00

Normal, Heavy Weight, and Mass Concrete, Practice for Selecting Proportions for, **ANSI/ACI 211.1-81,** $15.00

Packaged, Dry, Rapid-Hardening Cementitious Materials for Concrete Repairs, Specification for, **ANSI/ASTM C928-80,** ▲

Preparation of Notation for Concrete, **ANSI/ACI 104-71(R1977),** $6.00

Selecting Proportions for No-Slump Concrete, Practice for, **ANSI/ACI 211.3-75(REV80),** $10.00

Selecting Proportions for Structural Lightweight Concrete, Practice for, **ANSI/ACI 211.2-81,** $9.00

Shotcrete, Specification for Materials, Proportioning, and Application of, **ANSI/ACI 506.2-77,** $9.00

Shotcreting, Practice for, **ANSI/ACI 506-66(1978),** ▲

Shrinkage-Compensating Concrete, Recommended Practice for Use of, **ANSI/ACI 223-77,** $11.00

Shrinkage-Compensating Concrete, Test for Restrained Expansion of, **ANSI/ASTM C878-80,** ▲

Structural Concrete for Buildings, Specifications for, **ANSI/ACI 301-79,** $16.75

Structural Plain Concrete, Building Code Requirements for, **ANSI/ACI 322-71,** $6.00

Time of Setting of Concrete Mixtures by Penetration Resistance, Method of Test for, **ANSI/ASTM C403-80,** ▲

Unit Weight of Structural Lightweight Concrete, Test for, **ANSI/ASTM C567-80,** ▲

CONCRETE AGGREGATES

Evaluation of Frost Resistance of Coarse Aggregates in Air-Entrained Concrete by Critical Dilation Procedures, Recommended Practice for, **ANSI/ASTM C682-75(1980),** ▲

Lightweight Aggregates for Insulating Concrete, Specifications for, **ANSI/ASTM C332-80,** ▲

Petrographic Examination of Aggregates for Concrete, Practice for, **ANSI/ASTM C295-79,** ▲

Reducing Field Samples of Aggregate to Testing Size, Methods for, **ANSI/ASTM C702-80,** ▲

CONCRETE CONSTRUCTION

see also **Construction and Demolition**

Bins, Silos, and Bunkers for Storing Granular Materials, Recommended Practice for, Design and Construction of Concrete, **ANSI/ACI 313-77,** $16.00

Bonding to Hardened Concrete with a Multi-Component Epoxy Adhesive, Producing Skid-Resistant Surface with a Multi-Component Epoxy System, and Repairing with Epoxy Mortars (includes ANSI/ACI 503.1, 503.2, 503.3 and 503-4), **ANSI/ACI 503-79,** $15.00

Concrete Formwork, Recommended Practice for, **ANSI/ACI 347-78,** $15.25

Concrete Highway Bridge Deck Construction, Practice for, **ANSI/ACI 345-74,** ▲

Construction of Concrete Pavements and Concrete Bases, Practice for, **ANSI/ACI 316-74,** $12.00

†Forms for Two-Way Concrete Joist Floor and Roof Construction, **ANSI A48.2-1978,** $3.00

†Joint Construction, Forms for One-Way Concrete, **ANSI A48.1-1978,** $3.00

Nuclear Safety Related Concrete Structures, Code Requirements for, **ANSI/ACI 349-80,** $52.00

Piers, Construction of End Bearing Drilled, **ANSI/ACI 336.1-79,** $5.00

Reinforced Concrete, Building Code Requirements for, **ANSI/ACI 318-77,** $35.00

Reinforced Concrete, Building Code Requirements for (supplement to ANSI/ACI 318-77), **ANSI/ACI 318-80,** $7.00

Structures, Building Code Requirements for Concrete Masonry, **ANSI/ACI 531-79,** $17.00

Tolerances for Concrete Construction and Materials, **ANSI/ACI 117-81,** $5.00

CONCRETE CURING

see also **Concrete**

Atmospheric Pressure Steam Curing of Concrete, Practice for, **ANSI/ACI 517-70,** $14.00

Curing Concrete, Practice for, **ANSI/ACI 308-81,** $7.00

Water Retention by Concrete Curing Materials, Method of Test for, **ANSI/ASTM C156-80a,** ▲

CONCRETE JOINT SEALERS

Concrete Joint Sealer, Cold-Application Type, Specifications for, **ANSI/ASTM D1850-74(1979),** ▲

Concrete Joint Sealer, Hot-Poured Elastic Type, Specifications for, **ANSI/ASTM D1190-74(1980),** ▲

CONCRETE PRODUCTS

see also **Pipe, Concrete**

Water and Wastewater Structures, Precast Concrete, **ANSI/ASTM C913-79,** ▲

CONCRETE REINFORCEMENT

Details and Detailing of Concrete Reinforcement, **ANSI/ACI 315-80,** $7.00

Half Cell Potentials of Reinforcing Steel in Concrete, Test for, **ANSI/ASTM C876-80,** ▲

Steel Spirals for Reinforced Concrete Columns, **ANSI/SPR R53-63,** $2.00

Welded Steel Plain Bar or Rod Mats for Concrete Reinforcement, Specification for, **ANSI/ASTM A704-74(1980),** ▲

CONCRETE, INSULATING

see **Concrete; Concrete Aggregates**

CONCRETE, PORTLAND CEMENT

see also **Concrete**

Cement Content of Hardened Portland Cement Concrete, Method of Test for, **ANSI/ASTM C85-66(1973),** ▲

CONDUCTORS

see also **Wire and Cable, Electric**

Aluminum Conductor and ACSR, Packaging Standards for, **ANSI/AA 53-1981,** $5.00

Non-Specular Surface Finish on Bare Overhead Aluminum Conductors, **ANSI C7.69-1976(R1982),** $3.00

Temperature Rise as a Function of Current in Printed Conductors, Test to Determine the, **ANSI/EIA RS-251-A-1970,** $5.00

CONDUIT AND DUCTS

Duct and Fittings for Underground Installation, PVC and ABS Plastic Communications, **ANSI/NEMA TC10-1978,** $13.50

Duct for Underground Installation, Extra-Strength PVC Plastic Utilities, **ANSI/NEMA TC8-1978,** $5.50

Duct for Underground Installation, Fittings for ABS and PVC Plastic Utilities, **ANSI/NEMA TC9-1978,** $7.00

Duct, Corrugated Polyolefin Coilable Plastic Utilities, **ANSI/NEMA TC5-1978,** $5.00

Duct, Smooth-Wall Coilable Polyethylene Electrical Plastic, **ANSI/NEMA TC7-1978,** $5.00

Electrical Metallic Tubing, Safety Standard for, **ANSI/UL 797-1983,** $7.00

†Electrical Metallic Tubing, Zinc Coated, Specification for, **ANSI C80.3-1983,** $5.00

Fiber Conduit, Safety Standard for Electrical Impregnated, **ANSI/UL 543-1982,** $6.50

ALL PRICES SUBJECT TO CHANGE WITHOUT NOTICE

FIGURE 4.12 Example of page from ANSI Index of Standards.

As with other standards organizations, a system of letter prefixes has been established for the orderly grouping of subject matter for easier recovery of data

A. Source.
B. Treatment.
C. Distribution.
D. Storage.
E. Pumping.

ACI (American Concrete Institute)

This institute possibly represents the most respected of concrete standards in the United States. Most of the ACI provisions have been adopted in all other codes and regulations, with only minor changes. ACI is a nonprofit technical society. It is not industry-supported [Its nearest industry-supported counterpart is the Portland Cement Association (PCA).] ACI publications are produced by standing committees, each of which is identified by a committee number. All publications are identified by a numbering system in which the ACI committee number forms the publication number, followed by the revision date. Thus, committee 318 refers to the code committee, and each new revision of the ACI Building Code carries the same number 318, followed by the latest revision date.

There are numerous other standards that we have not mentioned, but the methods of identifying their respective publications are similar to those used by the organizations just mentioned. The following is a partial listing of some of the other organizations that publish standards frequently cited in public works specifications.

AASHTO	American Association of State Highway and Transportation Officials.
AISC	American Institute of Steel Construction.
AISI	American Iron and Steel Institute.
AITC	American Institute of Timber Construction.
APA	American Plywood Association.
APWA	American Public Works Association.

ASHRAE	American Society of Heating, Refrigerating, and Air Conditioning Engineers.
ASME	American Society of Mechanical Engineers.
AWPA	American Wood Preservers Association.
AWPI	American Wood Preservers Institute.
AWS	American Welding Society.
CRSI	Concrete Reinforcing Steel Institute.
IES	Illuminating Engineering Society.
NEMA	National Electrical Manufacturers Association.
NFPA	National Fire Protection Association.
SSPC	Steel Structures Painting Council.
UL	Underwriters Laboratories.
WIC	Woodwork Institute of California.
WRI	Wire Reinforcement Institute.

ACCESS TO SPECIAL STANDARDS BY THE CONTRACTOR

All too often, the specifier cites publications and standards that are not usually available in the field library of the contractor, or of anyone else, for that matter. There are companies that specialize in the sale of copies of such documents in many major cities in the United States; whose business it is to serve the needs of contractors. It often does little good to contact the resident engineer or inspector, as all too often they are not in possession of such documents either. Worse yet, many of the designers have never even read the standards that they cite in their specifications.

There are certain standards that have been so widely used for so long, however, that hardly anyone ever needs a copy of them for their personal use. Generally, all that is ever asked for by the architect-engineer is a manufacturer's certificate stating that the product is in conformance with a particular standard.

Some firms solve the problem in an ideal fashion. A volume made of copies of all the cited standards in a specific project specification is compiled and provided in the resident engineer's or inspector's field office at the beginning of the job. There, it is available for reference by anyone on the job whenever needed.

CHAPTER 4 SPECIFICATIONS AND DRAWINGS

BUILDING CODES, REGULATIONS, ORDINANCES, AND PERMITS

Building codes have been adopted by most cities, counties, and states. They are adopted by each governmental entity by ordinance or other means at their disposal to impart the force of law behind them. The codes carefully regulate the design, materials, and methods of construction, and compliance with all applicable code provisions is mandatory. Most of these codes are based, in whole or in part, on the various national codes that are sponsored by different national and regional groups. Changes, when made in the parent code, have generally been to accommodate local needs and conditions, and to make portions of the codes more stringent than may have been provided for in the code as it was originally written.

There are, of course, many codes that are not based upon such national codes, but have been specifically written for some particular locality. Many large cities do this by creating their own building codes.

In private work, the building department generally has jurisdiction over all construction and issues permits for construction after a review of the plans. In public works, however, there are often jurisdictional differences wherein a particular public agency is exempt from the control of the building department and may even construct buildings without obtaining a building permit from another department.

Typically, municipal engineering projects, such as those involving curbs and gutters, sidewalks, street construction and repair, storm drains, drainage structures, manholes, water or sewer lines, and similar construction, are performed by a different department of the public agency, and generally they do not fall under the jurisdiction of the building department, nor under the terms of the building codes insofar as they apply to the types of construction just named. Generally, however, any building or other habitable structure is subject to such codes and will be inspected by members of the appropriate department of building and safety.

There are several prominent codes in use in the United States. The more prominent codes are the following:

1. *National Building Code* Compiled by the American Insurance Association and adopted in various localities across the country.
2. *Uniform Building Code* Compiled by the International Conference of Building Officials (ICBO), a quasi-public organization regulated

BUILDING CODES, REGULATIONS, ORDINANCES, AND PERMITS

by the elected building officials of various cities and counties; widely used in the western states.

3. *Basic Building Code*   Compiled by the Building Officials and Code Administrators International (BOCA); used mainly in the eastern and north central states.

4. *Southern Standard Building Code*   Compiled by the Southern Building Code Congress (SBCC); used in most southern and southeastern states.

5. *National Electrical Code*   Compiled by the American Insurance Association; widely used in all parts of the United States.

6. *National Plumbing Code*   Compiled by the American Public Health Association and the American Society of Mechanical Engineers; used widely in all parts of the United States.

7. *Uniform Plumbing Code*   Compiled by the International Association of Plumbing and Mechanical Officials (IAPMO); widely used in the west.

8. *ICBO Plumbing Code*   Compiled by the International Conference of Building Officials (ICBO); used principally in some of the western states.

9. *ICBO Uniform Mechanical Code*   Compiled by the International Conference of Building Officials (ICBO); (Mostly HVAC and gas piping) used principally in the western states.

Of particular importance to the contractor is the "edition" date of the code. Even though a new code edition has been published, there can be no assurance that a particular jurisdiction will adopt it at any certain time, if at all. A case in point was a city in California, which for years used an old edition of the Uniform Building Code that it kept updated by adopting a series of city ordinances, annually, to meet the requirements of the local building official. Thus, in fact, the city's code was a special one, and only by carefully studying the basic, earlier edition plus all subsequent ordinances adopted could a contractor be aware of the conditions that affected its project in that jurisdiction. That particular city has since adopted a newer version of the Uniform Building Code; the cycle has begun all over again.

In addition to the provisions of any applicable codes, the contractor is obligated to conform to the provisions of all permits issued by public agen-

CHAPTER 4 SPECIFICATIONS AND DRAWINGS

cies having jurisdiction. Many of these permits are several pages long and resemble a small specifications book.

Projects Subject to Control by More Than One Agency

Often, a project will involve property or facilities that are under the jurisdiction of another agency. If a pipeline project, for example, crossed an interstate freeway route, a flood control right-of-way, or the pipeline of another public agency or utility, the contractor might well be faced with the prospect of having not only the project owner's inspector on the job, but a battery of other inspectors who represent each of the other affected agencies' interests. In each such case, a separate encroachment permit may be required from each affected agency, and the contractor would not only be bound by the terms of the contract documents for the project, but by the terms and conditions of each such permit as long as work was being done within the jurisdiction of the permit-issuing agency.

TYPES OF DRAWINGS COMPRISING THE CONSTRUCTION CONTRACT

Contract Drawings

The construction drawings or plans that were prepared especially for a given project are generally referred to as the *Contract Drawings,* and they have been prepared for the express purpose of delineating the architect-engineer's intentions concerning the project that it has conceived and designed. The drawings normally indicate the arrangement, dimensions, geometry, construction details, materials, and other information necessary for estimating and building the project. They do not necessarily show extensive small details but leave that up to the contractor to prepare in the form of shop drawings.

Standard Drawings

Occasionally, *Standard Drawings* of public agencies are defined as being a part of the contract documents. These drawings usually portray the repetitious details of certain types of construction that may be required by the local public agency for all similar work constructed within their jurisdiction. A typical example is the design of drop inlet structures and other drainage structures, or of standard curb and gutter designs that have been standardized in each community over a period of years. Thus, instead of redrawing the same details over and over, the architect-engineer simply refers to a certain Standard Drawing of the agency involved. A word to the wise, however; many such Standard Drawings contain outdated information, such as cast-in-place ladder rungs in manhole structures that do not meet OSHA standards, outdated ASTM references, or manhole frame and cover-casting numbers that are no longer stock items. The contractor must use extreme care in bidding these items while still adhering to the general rule that a contractor can only bid what is specified.

Shop Drawings

Shop drawings are those details and sketches prepared by the contractor or its subcontractors, fabricators, or materials suppliers that are necessary to assure the fabricator that its basic interpretation of the contract drawings is correct before beginning costly fabrication. Shop drawings frequently contain information that is not related to the design concept or that is relative only to the fabrication process or construction techniques in the field, all of which are normally outside the scope of the duties and responsibilities of the architect-engineer.

In approving shop drawings, the architect-engineer intends to indicate only whether or not the submitted drawings conform to the basic design concept of the project and are in general compliance with the terms and conditions of the Contract Documents. The contractor is made responsible for confirming dimensions in the field, for any information that pertains solely to the fabrication process or to the techniques of construction, and for the coordination of the work of all the trades.

Sometimes, an architect-engineer will attempt to make design changes directly on shop drawings and will indicate such changes in the form of "corrections" to the shop drawings. Check the shop drawings carefully after the architect-engineer returns them. You are within your contractual rights to demand a change order for any such work.

A change order is normally the only valid means of changing a contract after award. The proper way for an architect-engineer to make design changes after you have already submitted shop drawings for the previous design configuration is to return the shop drawings to you without action, with a statement that a change is in the making and a change order will be forthcoming. You will then have the opportunity to negotiate the cost of the change in terms of both time and money.

Keep in mind, however, that once you have submitted shop drawings, even though they may not be used due to the design change, you may be entitled to recover not only the cost of their preparation, but delay costs as well.

Record Drawings/As-Built Drawings

Some confusion seems to accompany the use of the terms *Record Drawings* and *As-Built drawings*. As-built drawings are a marked set of prints that are prepared by the contractor to show the differences between the details of the project as planned vs. how it was actually built. If you prepare as-built drawings, you are, in effect, certifying that the project was constructed exactly as you indicated on the as-built drawings. Often, an architect-engineer will be asked to prepare a set of tracings containing this same information. This is normally prepared from the drawings marked by the contractor and is called *Record Drawings*. The rationale here is that such drawings by the architect-engineer are a *record* of the information provided by the contractor.

All change orders should be reflected in the As-Built drawings, as well as any variances that may not be documented anywhere else.

The term *as-built* is unpopular with architects-engineers as such a declaration often places them in the position of certifying that the information provided by the contractor is accurate. On numerous occasions, such data have proved to be inaccurate or false, and the architect-engineer was on the losing end of an errors and omissions suit. The joint documents committee strongly advises against the use of the term by engineers.

CHAPTER · 5 ·

Construction Laws

- Compliance with Laws and Regulations
- Purpose of Construction Laws
- Limitations of Authority of a Public Agency
- Principal Types of Laws Governing Public Works Construction
- Public vs. Private Contracts
- Traffic Requirements During Construction
- Work within or Adjacent to Navigable Waterways
- Subcontracts
- Labor Laws
- Ethnic Minorities and Women in Construction
- Labor Relations
- Collective Bargaining
- Prejob Labor Considerations
- Open-shop Contracting
- Worker's Compensation and Employer Liability Insurance
- Licensing Laws

CHAPTER 5 CONSTRUCTION LAWS

It is safe to say that the entire area of public works construction is governed by laws. From the funding of a new public works project, such as an airport, to the design and construction and the final operation of the completed facility, every aspect is controlled and governed by some law, regulation, statute, or ordinance. By definition, a public works project involves the public interest, and nearly all involve public funds.

COMPLIANCE WITH LAWS AND REGULATIONS

Every public works construction project, whether at the local, state, or federal level, is governed by a set of laws administered by at least one jurisdiction. It is important for the contractor to know which laws apply to a given project. Those applicable laws identify and define the public agency's and the contractor's responsibilities under a given public works contract. The laws, regulations, statutes, and ordinances become the "rules" by which the public contracting game is played.

When a project is one being done by a single jurisdictional entity, such as a local city government, it is much easier to define the "rules of the game" than when multiple jurisdictional entities are involved. The following are just a few examples of the various jurisdictional levels that can apply to any given project:

1. Town council.
2. Unincorporated township.
3. Incorporated township.
4. City.
5. Municipal.
6. County.
7. State.
8. Federal.

It is not uncommon for a project to be subject to more than one set of jurisdictional laws, regulations, statutes, and ordinances. In the case of a municipal airport, for example, city, municipal, county, state, and federal laws, regulations, ordinances, and statutes could apply. Which jurisdictional agencies will be involved in a given project depends upon the ownership of the project, the location of the land being used for the project, the funding for the project, the purpose of the project, and ultimately who will control the project. In the example of the municipal airport, if it is to be run by the municipality or the city, or if it is to be built on county property with state funds and is to handle some interstate commercial airline traffic, all of the entities previously identified with the municipal airport would have some impact on the project.

Because of the infinite number of public works projects in terms of types, jurisdictions, funding arrangements, and applicable laws, an in-depth

CHAPTER 5   CONSTRUCTION LAWS

analysis of any specific public works project is beyond the scope of this book. Each project must be considered on a case-by-case basis, and it is hoped that the issues raised here will help the contractor understand the forces at work in a project.

Law in general and construction law applicable to public works projects in particular is continually changing and evolving. Changes in construction law occur as a result of court decisions and new legislation. Consequently, it is important to develop and maintain a program to keep abreast of the changes in the law that apply to the construction of public works projects.

PURPOSE OF CONSTRUCTION LAWS

The public's interest has historically been the driving force behind the evolution in construction laws applicable to public works projects. A public works project is one that by definition is for the benefit of the health, safety, and welfare of the public as a whole. Laws have developed to define the competitive bidding practices that are used throughout all levels of government in order to ensure the lowest possible cost to the public for construction of public projects. Such bidding laws, as well as other equal opportunity laws, have evolved in order to give as many contractors as possible an opportunity to bid on such public projects and receive public funds in payment. Some limitations apply, however, as in the case of small business set-asides, for which only those businesses qualifying as small businesses under the federal or state definitions are permitted to submit bids. Those, as well as other construction laws, have evolved to effect social change, such as equal opportunity, or to stimulate economic growth in certain geographical areas. The construction laws have also evolved in order to curtail or minimize corruption and the misuse of public funds.

Construction laws in public works contracts evolve when government entities institute programs to address specific problems. The Tennessee Valley Authority and the Environmental Protection Agency's (EPA) Clean Water Grant Program are two good examples. They are, also, good examples of how construction laws vary between projects depending upon the funding. The Tennessee Valley Authority project was basically a federal project completed with federal money by the Army Corps of Engineers, a federal agency. The EPA Clean Water Grant Program, on the other hand, involves local public agencies working on projects in accordance with federal and state guidelines in order to qualify for and receive federal funds. In addition to such defined programs as the Tennessee Valley Authority and the EPA Clean Water Grant Program, construction laws and programs may develop as a result of specific court decisions. As the issue of hazardous waste has developed, governmental entities and public funds have become more and more involved as courts have identified hazardous or dangerous conditions as those now requiring public attention and public funds, regardless of fault.

LIMITATIONS OF AUTHORITY OF A PUBLIC AGENCY

Although the contract entered into between the contractor and the public agency responsible for a given public works project will generally define the relationship between them, it is also important for the contractor to be aware of the limitations of the public agency's authority to enter into contracts in order to do certain acts. As previously stated, a governmental agency has no power to perform an act for which it is not properly authorized to do so. Under certain circumstances, that fact can have a major impact on a contractor.

A public agency has the ability only to do those things for which it has been granted authority. Whatever a public agency wishes to do by the way of a public works construction contract, the agency must be authorized to do so under the laws of that particular jurisdiction. A public entity cannot act in excess of its jurisdiction, and any action taken by a public agency without the legal power to do so is without legal force and effect. In other words, a public agency can do only what the law allows or prescribes. The contractor, by knowing the particular laws that apply when dealing with a public agency, can use that knowledge to its advantage in order to force the public agency to do what is required by law.

The way that a limitation of authority can ultimately affect a contract is shown in the case of *Zottman v. The City and County of San Francisco.* [(20 Cal. 96 (1862)] In that case, the City of San Francisco properly entered into a valid contract with a contractor to make improvements to a certain town square. After the contract was entered into, the scope of the contract was changed by the City, and the contractor was ordered to perform certain additional work. The contractor was paid for the contract work, but the City refused to pay for the extra work. The contractor then brought legal action against the City for payment. The court denied payment because the contract for extra work was not let to the lowest bidder after proper notification as required by law. The court stated that the acts of the City officials in demanding that the work be done did not ratify the contract for the extra work, as the City had no authority to make such a contract in the first place. Although the court acknowledged that denying payment to the contractor that performed the work might be a hardship, the court found that the contractor was aware of the law, and would suffer only what should have been anticipated. The City was allowed to retain the work without paying for it. According to later decisions, had the contractor received any payments for any portion of the extra work, the City would have been entitled to demand that the contractor return such progress payments to the City. The risks that a contractor takes in working with a public agency outside of the law or in violation of the law can easily be seen. In public works contracts a contractor is on constructive notice of an agency's jurisdiction and authority whether or not the contractor had actual knowledge. This is a factor that does not exist in private construction work.

PRINCIPAL TYPES OF LAWS GOVERNING PUBLIC WORKS CONSTRUCTION

As can be seen, the more jurisdictional agencies that are involved, the more laws there are to contend with. In general, however, construction laws fall into four major categories:

1. *Contract Law* Those specific laws and regulations affecting the making of contracts, both public and private, including case law that has developed through the court.
2. *Laws Governing the Execution of the Work* being performed under the contract, including the issuance and conformance to the various permits, regulations, ordinances, and other requirements of the various jurisdictional agencies that are involved.
3. *Laws Relating to the Settlement of Differences and Disputes* that may develop during the performance of the contract, including administrative remedies, case law, and quasi-judicial procedures such as arbitrations if applicable.
4. *Licensing Laws* that govern not only the business practices but also the qualifications of the various entities involved in the construction process, including contractors.

PRINCIPAL TYPES OF LAWS GOVERNING PUBLIC WORKS CONSTRUCTION

The last category includes the licensing of architects and professional engineers in every state in the United States, as well as the licensing of contractors in many states, of construction managers in a few states, and of inspectors in some states. In addition to such licenses, which require a demonstration of proficiency by some type of examination, there are local business licenses, permits to do business in certain areas, and sales tax permits. In addition, some jurisdictions even license certain building trades such as plumbers and electricians.

The more common laws encountered in most public works construction projects are the following:

> Davis–Bacon Act requirements (for federal or federally funded projects).
>
> Federal Acquisition Regulations (for contracts with the federal government) published in Book 48 of the Code of Federal Regulations (CFR48). The first two volumes are general and apply to all federal agencies. These are supplemented in separate volumes by additional regulations for certain specific federal agencies.
>
> OSHA safety requirements (both state and federal) for both industrial and construction safety. Federal requirements for safety are published in Book 29 of the Code of Federal Regulations. Industrial safety is in 29CFR1910 and construction safety is in 29CFR1926. Unless the local state requirements have been brought up to the federal standards, the federal regulations will apply to *all* work, whether federal money is involved or not.
>
> State Labor Code requirements.
>
> U.S. Department of Labor Requirements, as applicable.
>
> State housing laws.
>
> Local building codes and ordinances.
>
> Sales and use tax regulations.
>
> Air pollution control laws.
>
> Noise abatement ordinances.
>
> Business licenses to conduct business in each locality.
>
> Mechanic's lien laws of the state.
>
> Unemployment Insurance Code requirements.
>
> Worker's compensation laws.

Corps of Engineers and Coast Guard regulations for work in navigable waterways.

Subletting and subcontracting laws.

Licensing laws for architects, engineers, surveyors, contractors, and inspectors.

Construction permits by special local agencies, including the following and numerous others:

    Building permits.
    Demolition permits.
    Grading permits.
    Encroachment permits.
    Street work permits.
    Police permits for traffic control.
    Excavation permits.
    Environmental protection.
    Agency permits.
    Special hauling permits.
    Department of Agriculture permits.

Civil Rights Acts 1964 (federally assisted programs).

All local city and county or parish codes and ordinances.

Although there are many other relevant laws besides those listed, the purpose of this list is to call to the contractor's attention the fact that a public works project cannot be built without regard to the various laws, regulations, statutes, and ordinances of the many jurisdictional agencies affected by the work. These agencies include federal, state, county or parish, city, and special districts, along with many state and federal bureaus that have a legal interest in the effect that a project may have on the proposed areas that they have the legislative mandate to control. In addition, the list must include all others who have a legitimate interest in the work that affects facilities over which they have legal jurisdictional responsibility.

PUBLIC vs. PRIVATE CONTRACTS

A contract entered into between a contractor and a public agency includes all of the statutory law and case law applicable to private contracts in that jurisdiction in addition to all of the laws, regulations, statutes, and ordinances applicable to the particular project. Over a period of time, laws evolve requiring that applicable public works contracts contain the specific language of certain laws, regulations, statutes, ordinances, and other provisions in the Contract Documents. Most such provisions apply to public works contracts involving construction costs in excess of a certain statutory amount. Examples of such provisions are bidding procedures and requirements, fair subcontracting laws, identification of subcontractors, protection of public utilities, assignability of contracts, retainage, wage schedules, prevailing rates of wages, travel and subsistence, penalties for violating wage payments, payroll record requirements, equal opportunity employment requirements, recognized legal holidays, worker's compensation, shoring and bracing, trenching and permit requirements, "or-equal" products, relief of bidders, extra compensation, damages, and substitution of securities for retainage. In federal contracts or local public agency contracts in which federal funds are involved, additional provisions will generally be included related to audits, federal wage rates, equal employment opportunity provisions, and specific sets of federal regulations, such as the now superseded United States Environmental Protection Agency *Appendix C-2* (40CFR35 subpart (e), dated December 29, 1976). Many of the laws that are incorporated directly into the public works contract, as well as those that are not, are important to the contractor. These laws not only identify the public agency's rights and obligations with regard to third parties, they help to define the contractor's rights.

TRAFFIC REQUIREMENTS DURING CONSTRUCTION

When a project involves restrictions on local traffic in the area of the project, the contractor, unless specified otherwise in the Contract Documents, will be responsible for all traffic safety. When the project requires that the contractor close a street or intersection, or restrict its traffic, the contractor before doing this must obtain a permit from the appropriate local agency.

Generally, the terms and conditions of any such permit will become a part of the contract provisions. It then becomes the contractor's obligation to ensure the safety and welfare of the public and to observe the traffic requirements as required by the permit. Some local agencies have developed their own manual in order to assist contractors working on projects that affect the traffic within a given jurisdiction; others rely upon the Federal Highway Administration guidelines and often include its provisions as a part of the Contract Documents by reference. [ANSI D6.1-1978 Manual on Uniform Traffic Control Devices for Streets and Highways, Part VI: Traffic Controls for Street and Highway Construction and Maintenance Operations; U.S. Department of Transportation, Federal Highway Administration]

Although there is no question that the contractor must furnish proper traffic controls in order to protect the health, safety, and welfare of the public, a contractor must carefully consider what kind of controls will be required in order to properly prepare its bid. A particular jurisdiction may require that certain excavations be continually fenced or covered; that not more than a specified length of pipe trench be allowed open at any one time ahead of or behind a pipeline operation; or that a contractor maintain safe access to certain businesses and residences by providing temporary "bridges" over trenches or other excavations until such excavations have been properly filled and compacted. Unless the contractor is well aware of these requirements during the preparation of its bid, it may substantially underestimate the cost of performing the work. For example, when a new section of highway is being constructed that includes a railroad crossing, the contractor may be required to control traffic over the railway tracks as well as at the highway connections. The contractor may carefully calculate the cost of traffic control measures related to the highway work and include this information in its bid while forgetting to investigate the railroad requirements for traffic control measures, which will also be the responsibility of the contractor. In one particular case, a contractor had done just that and wound up having to spend considerably more money than it had included in the bid to protect the railroad crossing with contractor-paid flaggers. Railroad crossings require the contractor to work out certain details with the railroad involved, and traditionally railroad crossings within a project area have been the source of numerous construction delays and costs not anticipated by the contractor when making up its bid.

Traffic control requirements are also important to the contractor from the standpoint of liability. Whenever a construction project is underway, the project itself, along with the areas surrounding the project, are in a state of flux. The condition of the jobsite and the surrounding area has changed from what a traveler familiar with the area has come to expect, and due to its changing nature, the work increases the potential for accidents. The contractor should take every step possible to minimize the chance of an accident occurring.

When an accident does occur, however, it often results in property damage and personal injury for which the contractor will be held liable. It is a well-established principle that a contractor engaged in building or repairing a road owes a duty to exercise ordinary care to protect motorists from injury as a result of the construction work. The exercise of such care requires the erection of guards and the placing of lights or barriers at obstructions, and the violation of any ordinance requiring such lights or barriers would constitute negligence per se. As public works contracts always require a contractor to be insured, such claims would be covered, less any deductible amount, by insurance. However, as with all insurance, any claims against the contractor's insurance policy will ultimately raise the contractor's insurance premiums. Insurance costs should be specifically and thoroughly investigated prior to submitting a bid. The recent trend in insurance coverage is to limit the carrier's liability under the policy to an amount considerably under that previously available, in spite of the fact that the value of liability claims is increasing by leaps and bounds.

WORK WITHIN OR ADJACENT TO NAVIGABLE WATERWAYS

The Rivers and Harbors Act of 1899 is alive and well and seems ready to outlive all of us. When a project involves work within or adjacent to "navigable waterways," laws applying to such waterways become applicable to construction contracts that involve such waterways. All work being constructed in or involving the use of such navigable waterways is primarily subject to the orders and regulations of the Department of the Army, Corps of Engineers (see Figure 5.1). Insofar as the movement of a vessel or traffic control on such waterways is concerned, the U.S. Coast Guard also shares jurisdiction.

The term navigable waters is much broader now than in previous years, and is administratively defined in *Permits for Activities in Navigable Waters or Ocean Waters*, page 31324 of the Federal Register of 25 July 1975. Watch out for this definition. In many western states, streams are included in the category of navigable waterways even though they may be bone dry six months out of the year. During the course of the work, a contractor is responsible for anything deposited or falling into a body of navigable waters that could be interpreted as a potential hazard to navigation. The contractor must give notice to the inspector of any such obstruction, along with a description and location of it. If necessary, the contractor may be required to place a buoy or other marker at the location of the obstruction until its removal.

CHAPTER 5 CONSTRUCTION LAWS

FIGURE 5.1 Offshore work, such as this 8850-ft ocean outfall, is subject to Corps of Engineers jurisdiction.

The contractor should be aware that for offshore construction, it is not uncommon for the terms of the contract documents to require that the contractor furnish the engineer with a two-way radio that is capable of communication with the work vessels.

The cost of conducting construction operations within or adjacent to navigable waterways, including the cost to identify, locate, and remove obstructions placed or lost in the navigable waterway, is the responsibility of the contractor unless otherwise stated in the Contract Documents.

In addition, the Federal Water Pollution Control Acts Amendments of 1972 (Public Law 92-500) requires a Corps of Engineers permit, under Section 404 of that Act, for the discharge of any more than one cubic yard of dredged or filled material into any navigable waters. A word of caution is appropriate here: The list of waterways defined as navigable waterways under this Act is different from those defined in the Rivers and Harbors Act of 1899. Stripped of its regulatory language, the "404 Permit," as it is commonly referred to in the industry, states that whenever a party deposits more than one cubic yard of anything in, on, or encroaching on any portion of the historic high water line of any "navigable waterway," a Corps of Engineers "404 Permit" is required. This, as interpreted by the District Engineer of one of the major Corps of Engineers Districts from court determinations, includes portions of dry land that have carried flood flows and to previous stream beds even where, after a flood, a stream followed a new watercourse after the flood. Take into account, also, that the lead time in obtaining a "404 Permit" is, at the very least, several months. Generally, it is preferable to plan construction operations to avoid entirely such encroachment or to apply for several permits in case one is disallowed.

A contractor must take the cost of such operations and the time and monies required to obtain necessary permits into account at the time it prepares its bid. Obviously, the contractor will also be liable for damages suffered by a vessel as a result of a contractor's negligence.

SUBCONTRACTS

In the evolution of construction laws on public works projects, the requirements for competitive bidding on all such contracts have probably been the most important. In preparing a competitive bid, a general contractor reviews the plans and specifications and asks the appropriate specialty contractors and minority, women-owned, and otherwise disadvantaged contractors to submit subcontract bids on their respective items of work contained in the project. A general contractor will then add the quotations received from the subcontractors to the amount estimated for the general contractor's portion of the work, plus the addition of an appropriate amount to cover overhead and profit, which would then total the contractor's bid price. If the general contractor's bid is the lowest, it would be awarded the job by the public agency, and the public would expect to have the project built for the lowest possible cost.

The general contractor moving into the public sector should be aware that some public construction contracts limit the proportion of the total amount that a prime contractor may subcontract. The Public Buildings Service of the General Services Administration, the U.S. Corps of Engineers, as well as other federal agencies and state agencies such as the California Department of Transportation have set such limitations. In addition, some states have enacted laws that impose somewhat similar restrictions on the general contractors of all public works.

Although such laws are intended to prevent the general contractor from subcontracting in excess of a specified proportion of the work on a project, the requirement can have a substantial impact on a general contractor. It can prevent the effective scheduling of job operations, lead to division of authority, create problems in the coordination of construction activities, weaken communications between management in the field, spawn jurisdictional disputes between unions, fragment responsibility, and reduce job efficiency. If such laws are applicable in an area where the contractor must work, they must be taken into account when preparing the bid.

Bid Shopping or Bid Peddling

In some geographical areas, once the general contractor is awarded a job, there is no obligation to use any of the subcontract bids that were used in computing the general contractor's bid. A general contractor, knowing already the amount of money that it will receive for each portion of the work, would then contract various other specialty contractors, including the one that tendered the original subbid, in an attempt to obtain a better price and reduce the amount that the general contractor would have to pay for that portion of the work, thus increasing significantly the profit potential. This practice is referred to as "bid shopping" or "bid peddling."

In order to protect subcontractors and the public against the practice known as "bid shopping" or "bid peddling" by general contractors, laws have been enacted in various parts of the United States to curtail such practices on public works projects. Although bid shopping is still very much a part of private contracting, fair subcontracting laws limiting the practice in public works projects have been enacted in some areas in order to assure the public of the full benefits of fair competition among private contractors and subcontractors, to ensure the integrity of the competitive bidding process. Such laws were enacted in response to the fact that bid-shopping practices often resulted in poor quality material and workmanship to the detriment of the public and led to insolvencies, loss of wages to employees, and other problems.

However, these laws can work to the benefit of the contractor. Under some of these laws, a contractor cannot be forced by a public agency to substitute or change subcontractors simply because the agency has no confidence in a named subcontractor or because it has had a previous unsatisfactory experience with one of the named subcontractors.

Effect of Bid Shopping

Beyond the immediate effect that bid shopping can have on individual contractors and subcontractors, the practice can affect the bidding process in a number of significant ways. It creates even more of an adversarial relationship between contractors and subcontractors; it decreases the accuracy and reliability of bids at all levels; it encourages general contractors to underestimate their bids based upon the assumption that offers from subcontractors can be lowered later by bid shopping; and it encourages subcontractors to inflate their bids in anticipation of the general contractor's bid shopping after the contract is awarded.

Over the long term, bid shopping in a particular area so reduces subcontractors' margins that they simply go out of business. Subcontractor competition is reduced and those remaining are free to raise their prices at will. At its worst, bid shopping can cause some contractors to submit irresponsibly low bids that result in the subcontractor sacrificing quality or skimping on safety or inspection procedures to the detriment of the project and the public. Subcontractors, after lowering their bid too far, very often go broke in the middle of a project. Such business failures ultimately lead to increased costs and reduced quality. In some cases, a subcontractor merely adjusts its bid to the general contractor according to its past experience with certain general contractors.

Often overlooked is the fact that bids solicited through the bid-shopping process force a subcontractor to limit its participation in the contract to its strict interpretation of the scope of its subcontract. This increases the prime contractor's contract administration costs and results in added costs of negotiation of disputes and attempts on the part of a prime contractor to impose its own interpretation of the subcontract terms, provisions, and scope.

Bid shopping has been condemned by the Code of Ethical Conduct of the Associated General Contractors of America for more than 30 years.

Control of Bid Shopping

At one time, in order to control bid shopping, "bid depositories" were created and managed by trade associations. A number of subcontractors would submit their sealed bids directly to the depository, which subsequently opened, tabulated, and printed all the bids on a specified day. Because of antitrust litigation, bid depositories have disappeared. There are other shortcomings to this system as well, as it is unsuited to a bidding environment where the general contractor does not do a significant amount of work with its own forces, but "brokers" the entire job.

Protective Legislation for Subcontractors

State legislation in some states has enacted "bid listing" statutes that require general contractors to provide the names of all subcontractors whose bids were used in preparing the general contractor's bid. Such statutory schemes also provide limited circumstances under which subcontractors listed by the general contractor can be changed. Some states and many federal construction contracts require subcontractor listing. For example, the regulations of the General Services Administration require a general contractor to list all subcontracts on prime bids. The Comptroller General has enforced this requirement.

Under another recently developed scheme, some state laws require general contractors to file all solicited bids of subcontractors with the state-awarding agency. The agency then rejects nonresponsive bids and forwards a list of acceptable subcontractor bids, based on price and other objective criteria, to the general contractor. The general contractor must then work from the list of acceptable subcontractors and prepares its prime bid in accordance with the subbids chosen. The substitution of subcontractors by the general contractor after the award of a contract can only be done with the approval of the public agency, and any substituted subcontractor must come from the agency's approved list. Connecticut and Massachusetts have enacted such "bid filing" laws, and other states may soon follow.

A more unique approach has been adopted in North Carolina. Subcontracts for the performance of specific categories of specialty work such as heating, ventilating, and air conditioning; plumbing; and electrical work on state-financed projects are awarded separately by the state-awarding agency, which eliminates altogether the possibility of bid shopping.

Under the California approach, which applies to all public works contracts, not just those at the state level, the law requires the prime contractor to set forth in its bid or offer the name and location of each subcontractor that will perform work and render services on the project in an amount in excess of one half of one percent of the total of the prime contractor's bid. The prime contractor, under such laws, must set out the portion of work that will be done by each named subcontractor, and only one subcontractor may be named for each such portion of the work.

Under the provisions of the California law, if the general contractor does not list a subcontractor for a particular portion of the work, the general contractor must perform the work with its own work force. A cash penalty of ten percent of the subcontract bid may be assessed by the public agency against a general contractor that fails to list a subcontractor at the time of bidding and then attempts to name a subcontractor after the award of a contract. This can often cancel any benefits that could otherwise have been obtained through the bid-shopping practice. In recognition of the fact that unforeseen events do occur, the California law does contain a provision to allow the substitution of a listed subcontractor with an unlisted subcontractor in a specific manner and under certain conditions. Generally, however, this is limited to cases in which a subcontractor is insolvent, unlicensed, or has been on the job and fails to perform.

LABOR LAWS

A contractor who, as an employer, hires workers is subject to a large number of federal and state statutes. What statutes apply to the contractor in any given situation is dependent upon the type of job, be it local, state, or federal; the state where the job is located; and the specific labor practices in the locale of the project. It is important that the contractor be aware of the laws that apply to its projects as they are crucial to the contractor's ability to function in today's labor environment. The labor movement within the United States is in large part responsible for the various labor laws that have been enacted throughout this country and, as such, labor laws and labor relations will be discussed here together.

The Sherman Anti-Trust Act

The Sherman Anti-Trust Act marked the beginning of federal laws relating to labor and management policies. Under federal law today, workers have the right to form and join unions and to take collective action to improve their economic condition. The Norris–LaGuardia Act of 1932 protected the rights of workers who strike and picket peacefully, while limiting the power of the courts to issue injunctions against union activity in labor disputes. Subsequently, states passed similar laws. The National Labor Relations Act of 1935, also known as the Wagner Act, was passed to protect union organizing activities and assist in collective bargaining. Under the Act, employers could not discriminate against employees for labor activities. In 1947, the Act was amended by the Taft–Hartley Act, which placed controls on the activities of organized labor. The Act established that every worker had a right to participate in or refrain from union activities.

The Landrum–Griffin Act

The Labor Management Reporting and Disclosure Act of 1959, the Landrum–Griffin Act, imposed controls on internal union affairs to protect the rights of union members, insure democratic elections, and fight corruption and racketeering in unions. The act was designed to protect the public against unscrupulous union activity.

The Civil Rights Act of 1964 set forth basic individual rights pertaining to voting, access to public accommodations, public utilities, public education, participation in federally assisted programs, and opportunities for employment. Title VII of the Act, the Equal Employment Opportunity section, prohibits discrimination in employment. The administration and enforcement of this law rest with the Equal Employment Opportunity Commission (EEOC), which has the responsibility to insure that hiring and promotion considerations are based solely upon the ability and qualifications of the individual considered.

The Equal Opportunity Employment Act

The Equal Opportunity Employment Act of 1972 amended the 1964 Act. The 1972 Act extended coverage to apprenticeship and other training programs and to employers with 15 or more employees.

LABOR LAWS

Executive Order 11246

Executive Order 11246, issued in 1965, concerned contracts and subcontracts exceeding $10,000 on federal and federally assisted (funded) construction projects. Contractors are prohibited under this order from discriminating against any employee applicant because of race, color, creed, or national origin. Each federal agency is responsible for insuring compliance with the order. A contractor is required to provide information showing that it is in compliance with the nondiscriminatory requirement of its contract, including the affirmative action clauses. When there is noncompliance, contracts may be canceled or suspended and the offending contractor declared ineligible for future government or federally assisted construction contracts. A contractor's compliance with these provisions is absolutely essential as the *agency has the authority to withhold progress payments* from contractors in violation of the order.

The Philadelphia Plan

In 1969, the United States Department of Labor imposed the Philadelphia Plan. Under the plan, contractors on federally assisted and federal construction contracts in the Philadelphia area were required to follow a standard specified range for minority group employment in the highest paid construction trades. Along with any bid, the contractor was required to include an Affirmative Action Program setting forth acceptable goals for the use of minority workers in each trade.

Similar plans were developed for most major metropolitan areas and are still in effect.

The Davis–Bacon Act

The Davis–Bacon Act of 1931 [40 USC §276(a), et seq.] and the federal regulations implementing it [20CFR Part 1; 45CFR Parts 100(a), 100(b)] require that wage rates, including fringe benefits, determined in advance must be paid to all workers on federally assisted projects. The Act has been amended to include a variety of public works projects that are financed in some way by the federal government. The concept of "prevailing wage" now applies to construction projects involving interstate highway systems, federally assisted schools, slum clearance, urban rehabilitation, hospitals, pollution abatement programs, airports, and certain types of FHA housing.

The Act is administered by the U.S. Department of Labor. Contractors with projects that are covered by the Act must keep records and make periodic reports showing compliance with the various regulations on the use of apprentices. Simply stated, the Act mandates that wage rates shall not be less than the prevailing wages as determined by the Secretary of Labor and published in the Federal Register for similar work on similar projects in the area where the project is being constructed. Under the Act, contractors are required to pay employees once each week and make only those deductions provided for under the Copeland Act.

Many states have enacted their own versions of the Davis–Bacon Act that extend similar requirements to all public projects within their jurisdictions, regardless of whether federal money is involved. On federal projects in such states, in the case of conflict between the two wage rate schedules, the contractor is obligated to pay the higher of the two.

The hidden cost here is that if a contractor has a project of long duration, new wage rate determinations for the project area issued after the work has begun will often contain higher wage scales. The law only requires a contractor to pay what was determined at the time the construction contract was signed. However, the contractor's labor force may not be willing to accept the original rates after new labor rates have been published by the Department of Labor.

The Copeland Act

The Copeland Act of 1934 makes it a punishable offense for an employer to deprive any individual employed on a federal construction project or a project financed in part or in whole by federal funds of any part of the compensation to which the worker is entitled. It is the purpose of the Act to prevent an employer from taking kickbacks from its employees, as a course of business or by force, intimidation, or threat of dismissal, and so on. Under the Act, payroll records must be maintained and reports submitted as required by the Department of Labor. This Act applies to all projects under the Davis–Bacon Act. In 1938, Congress enacted the Fair Labor Standards Act that determined minimum wages, maximum hours, overtime pay, equal pay, and child labor standards. The Act provides for a minimum wage for all employees covered by it and the required payment of overtime at the rate of one and one-half times the regular hourly pay for all hours worked in excess of 40 hours in one week.

The Hobbs Act

The Hobbs Act, which is also known as the "Anti-Racketeering Act of 1946," made it a felony to obstruct, delay, or affect interstate commerce by bribery or extortion. The Act was intended to prevent unions from demanding payoffs from employers in order to avoid labor trouble.

The National Apprenticeship Act

The National Apprenticeship Act of 1937 authorized the establishment of a Bureau of Apprenticeship and Training by the U.S. Department of Labor. The purpose of the Act was to create and improve apprentice programs. Federal regulations have been enacted to increase minority group participation in apprentice programs by requiring unions and contractors to take affirmative steps beyond nondiscrimination. The objective is to increase the skills of construction workers and the number of minorities in the apprenticeship programs.

ETHNIC MINORITIES AND WOMEN IN CONSTRUCTION

Affirmative action to increase the participation of women and ethnic minorities in the construction force came about with the Public Works Employment Act of 1977, which made available $4 billion of federal subsidies for construction projects through local governments, but required at least 10 percent of the work to be awarded to ethnic minority companies (MBEs), subcontractors, subcraftsmen, or suppliers. When bidding on a project involving such funds, the contractor must be aware of what is required in order to comply with the Equal Employment Opportunity (EEO) and Minority Business Enterprise (MBE) provisions. In many circumstances, a woman-owner business (WBE) qualifies as an MBE.

Although a history and full explanation of the effect of affirmative action programs such as these on construction is beyond the scope of this book, it is sufficient to say that bidding on many projects depends upon complete compliance with all such requirements. Particular attention should be paid to the sources of funding being used on any project to ensure that the contractor is aware of any such program requirements. If they are applicable, a contractor must scrupulously comply with each and every requirement in order to prepare a responsive bid and to stand a chance of being awarded the contract. Remember the Golden Rule: "He who has the gold makes the rules." Thus, the source of construction funds is an important issue to the bidder.

LABOR RELATIONS

Although the strength of union membership varies with geographical areas, all contractors, whether union or nonunion, are influenced by union labor policies and work rules in their specific areas. Unions have helped organize and stabilize the operation of the construction industry through negotiated labor contracts and the establishment of wage rates. In addition to providing certainty of labor costs, unions provide labor resources from which contractors can obtain the necessary qualified workers.

Construction unions are made up of *locals* that are given jurisdiction over designated geographical areas, such as a city, county, or state. A local has jurisdiction over a union worker in a craft within that area. Each union local elects its own officers and takes care of the day-to-day affairs of the local. The local is represented on each job by a steward who monitors the job for compliance with union rules. Locals group together to form Divisions, and ultimately different construction unions unite to form building and construction trade councils. During collective bargaining, delegations from each member local council represent a united front to the employers. The contracts and rules that result from such collective bargaining vary from area to area.

Due in part to the high and ever-increasing construction labor costs, and in part to the changing attitude of the public in general toward labor, union influence is changing. In some areas, unions are easing work rules. The enactment by some states of right-to-work legislation as well as the increased practice of open-shop contracting are signs that labor unions in the construction industry are in a period of adjustment.

COLLECTIVE BARGAINING

The National Labor Relations Act requires management and labor to bargain in good faith with one another. It does not require that concessions be made by either party or that the two sides come to any agreement. Labor negotiations in the construction industry either involve associations of contractors or individual contractors bargaining with a union or unions. The resulting contracts arrived at through such collective bargaining are usually local in nature. Only a few unions negotiate on a national basis.

Contractor associations are in a better position to negotiate with labor than an individual contractor, but they negotiate for the group, not for the specific needs of one company. Contractors can participate in the collective bargaining process through a local branch of the Associated General Contractors of America (AGC). National negotiations are handled by a committee of contractors, legal counsel, and AGC staff; unions are represented by officers of locals, district councils, and building trade councils. The end-product of such negotiations is a written agreement called a *labor contract* or *labor agreement,* which is a binding legal document for the term of the contract. In addition to setting forth wages, hours, holidays, and the like, the contract covers fringe benefits, including such items as pensions, hospitalization insurance, profit-sharing plans, paid vacations, sick leave, unemployment payment plans, and so forth, as well as overtime rates, premiums for high quality work, travel time, assistance, apprenticeship, and cost-of-living policy.

PREJOB LABOR CONSIDERATIONS

If conditions on a particular project are not of a usual nature, a contractor should meet with union representatives, if applicable, to identify any additional labor costs that may be involved in the project. For example, if the job is in an extremely remote area, consideration must be given to quarters, subsistence, transportation, and travel. If those items are not addressed early on and before the bid is prepared, they could result in major unforeseen costs to the contractor or, worse yet, a disruptive labor dispute.

Another consideration of a job located in a remote area is the ability to locate and attract a qualified work force to do the job. During meetings with local union representatives, arrangements to bring in additional workers from the outside, either by the union local or the contractor, can be discussed. Although many contractors may want to resist any prejob labor conferences for fear that labor will take advantage of such a situation, a good prejob labor conference will result in benefits to both sides. Sometimes when insufficient union labor is available to build an entire project, a local building council might allow open-shop conditions to exist on that particular federal project. By allowing both union and nonunion subcontractors on the same project, the trades council was able to ensure jobs for its members for the benefit of the community. In the process, the needs of the contractor are addressed. By law, federal projects must permit both union and nonunion subcontractors on the same job if so arranged by the contractor. This often gives rise to labor disputes, however, as many unions object strongly to working side by side with nonunion employees on the same project. As a result, a picket line may be set up, disrupting the contractor's work effort. This can be somewhat minimized, however, by invoking the provisions of the Taft–Hartley law that restricts picketing to a separate gate to the project site.

OPEN-SHOP CONTRACTING

Open-shop contracting means contracting unhampered by union agreements or representation. Generally, open-shop contracting occurs in areas where, for whatever reason, unions are not as strong. The most recent trend in the construction industry has been to move away from unions and more and more to embrace the open-shop concept. Some sources now estimate that close to 40 percent of the annual construction volume in the United States is now done by nonunion contractors. Many contractors, both small and large, are electing to go nonunion. Some contractors have two divisions in their organizational structure: one union and the other nonunion.

Although open-shop contractors pay somewhat less than union scale, they are not necessarily antiunion. Many contractors have been forced to become open shop or to go nonunion in order to survive in the face of recent construction industry economics. Open-shop contractors have the right to decide on the size of a work crew and to what job a worker may be assigned. Very often, such shops have a mixture of union and nonunion help, depending upon the circumstances. Workers are paid according to the work involved and their performance. However, an open-shop contractor bidding a federal job is subject to the Davis–Bacon Act and must pay all employees the minimum wage rates established by the Department of Labor, and the contractor's bid will have to be computed accordingly. Federal law does not prohibit an open-shop contractor from being awarded a contract on a federal job. Whether union or nonunion, all contractors on federal jobs or federally funded jobs must pay the same minimum rates of wages as determined by the Department of Labor.

WORKER'S COMPENSATION AND EMPLOYER LIABILITY INSURANCE

The contractor's Comprehensive General Liability Insurance policy normally excludes coverage of any liability incurred under state Worker's Compensation laws. The laws of every state require employers to be responsible for the payment of compensation benefits to employees who sustain job-related illnesses or injuries. Contractors normally finance such payments through the purchase of Worker's Compensation Insurance. Under worker compensation schemes, the insurance carrier then pays benefits to an injured employee at rates set by state law.

Very often, Worker's Compensation Insurance is frequently the most expensive type of insurance that a contractor must secure. There are numerous reasons for the high cost: the hazardous nature of construction work in general and the many job-related accidents that occur, long duration of payments, the statewide nature of Worker's Compensation programs, the pro-claimants on the Worker's Compensation boards, and the ease and simplicity of filing a claim.

Although some states require contractors to purchase Worker's Compensation Insurance from state-administered insurance trust funds, contractors in most other states are free to purchase such insurance from private insurance carriers. There are significant variations in the administration and levels of benefits paid out from state to state that can pose a problem for contractors doing business in more than one state. When a contractor does have operations in more than one state, private carriers often provide "all states" endorsement to cover the contractor for every state listed by the contractor in the policy. To ensure continued Worker's Compensation Insurance coverage, the contractor must supply updated information to the carrier, as required.

The cost of premiums for Worker's Compensation Insurance varies greatly as they are geared, in part, to the contractor's safety record. If a contractor has a good record of jobsite safety, it will pay very small premiums in comparison to those paid by a contractor with a consistent loss record. The opportunity for a safety-minded contractor with a good jobsite safety record to save money on reduced premiums is significant. When purchasing such insurance, a contractor should seek a policy from a carrier that will provide comprehensive loss prevention services as well as standard claim services.

LICENSING LAWS

The contractor may be required to obtain a number of licenses, the most common being the business license as required by the various local jurisdictions for the contractor to do business. In addition, some states require contractors to obtain a license as a contractor. Some state-licensing schemes call for different licenses for various types of contractors and require a contractor to pass a written examination before being issued a license. Some states require that a contractor post a bond with the state before the contractor's license becomes effective. Such bonds are in the control of the state agency and may be subject to claims against the contractor.

If a state requires a contractor to be licensed, any contractor who is unlicensed and contracts for work within that state may not only be subject to the penalties imposed by the state itself, but may have the validity of any contract it enters into challenged. When a license is required, state law often prohibits a contractor from seeking compensation for work done without a contractor's license. In some states, a contractor is required to have been in business as a contractor for up to six months before being permitted to bid on public works projects in that state.

When entering into a new jurisdiction, it is particularly important that the contractor be aware of any licensing requirements that may restrict a contractor's ability to operate. For guidance, a contractor may want to contact a local contractor's association, such as a local chapter of the AGC or ABC.

CHAPTER 6

Construction Safety

- Basic Responsibility for On-Site Safety
- Primary Liability
- OSHA and the Contractor
- Contractual Safety Requirements
- Accident Records
- Protection of the Public
- Trench and Excavation Safety

BASIC RESPONSIBILITY FOR ON-SITE SAFETY

Under the terms of most, if not all, public works construction contracts, the contractor is totally responsible for the project. That responsibility includes every aspect of project safety. Construction work is hazardous, and the frequency of accidents in the construction industry is high. Therefore, it is important that the contractor address the hazards associated with construction operations and implement an accident-prevention program to minimize accidents with the goal of obtaining an accident-free jobsite.

Some sources have estimated, without taking into account the consequential losses, that construction-related accidents cost more than $3 billion a year. Insurance is available to protect the contractor from direct expense related to construction accidents. Worker's Compensation provides hospital and medical care, as well as other types of payments to workers and their families; however, many other additional costs are never covered by any type of construction insurance. Construction accidents can cause damage to work already in place or in progress, increased administrative and legal expense, a reduction in employee morale, down-time and delay of the project, increased insurance premiums, and damage to a good reputation. Although such losses are somewhat intangible, they are nevertheless real and are not covered by any form of insurance.

Legislation relating to the regulation of working conditions was first passed in Massachusetts in 1867. From that time forward, other states have followed with legislation pertaining to various types of working conditions. In 1911, Wisconsin was one of the first states to establish a State Industrial Commission that was charged with promoting health and safety in industry.

Although safety codes differ from state to state, all states have regulations pertaining to potentially dangerous devices, such as elevators and boilers. Other regulations address such items as first-aid equipment required on certain projects, types of protective clothing, protective breathing apparatus, material-handling requirements, and so forth.

PRIMARY LIABILITY

Although the general contractor is ultimately liable for safety on a construction project, each one of the subcontractors should be made responsible for the health and safety of its own employees. The subcontractors, like the general contractor, will be held liable if their employees are exposed to hazards that could cause serious physical harm or death. Often, the general contractor is held liable for subcontractor-caused conditions and accidents.

Subcontractor Involvement with Safety

When safety requirements are specified in the prime contract, they should definitely be included in all subcontracts. If contained in the prime contract, and not in a subcontract, the contractor may find its progress payments held up for the safety violations of its subcontractor without any contractual basis to demand that the subcontractor bring its operations into compliance. Only by coordinating the safety requirements in the subcontracts with those in the prime contract can the contractor prevent this from happening.

With a contractor's potential liability as great as it is, it is probably a good idea to include safety requirements in all subcontracts, even if not specified in the prime contract. The safety requirements become a part of the contractual obligations of the subcontractor and give the general contractor a means to assure compliance with them.

Subcontractors Should Be Monitored

As the general contractor may ultimately be held liable for any accident that occurs on the construction site, careful attention must be paid to ensure that the subcontractors require their workers to follow all applicable safety and health regulations; supply their workers with any required personal protective equipment; and avoid the use of unsafe equipment on the job.

To ensure that its subcontractors, who very often follow voluntarily safety and health standards, do in fact comply with all such requirements, the contractor should consider including safety and health requirements in the terms of its contracts with subcontractors.

OSHA AND THE CONTRACTOR

In 1970, Congress found that personal injuries and illnesses arising out of work situations imposed a substantial burden upon, and were a hindrance to, interstate commerce in terms of lost production, wage loss, medical expenses, and disability compensation payments, and accordingly it passed the

CHAPTER 6 CONSTRUCTION SAFETY

Williams–Steiger Occupational Safety and Health Act of 1970 (OSHA) [Title 29 USC 451, et seq.]. With the passage of this Act, the federal government imposed nationwide safety standards on the construction industry. Under the Act, each state was allowed to pass its own version of OSHA, so long as the state's plan was at least as strict as the federal standards. OSHA imposes strict employee safety and health standards to protect covered employees and enforces the same provisions for inspections, investigations, record-keeping requirements, and enforcement procedures. Under OSHA, logs of accidents as well as supplementary information and inspections may be involved.

If a state exercises its right under the Act to enact a safety plan at least equivalent to the federal OSHA regulations, it retains the right to be the sole safety enforcement agency within that jurisdiction. If a state does not come up with such a plan, construction in that state will be subject to inspection by federal safety inspection agencies as well as state inspection agencies.

The Act established the Occupational Safety and Health Administration, U.S. Department of Labor, Washington, D.C. 20210. Regional offices are found in various cities throughout the country. OSHA is responsible for the establishment of safety and health standards and for the rules and regulations to implement them. Such rules and regulations are published in the Federal Register and can be obtained from local OSHA offices or federal bookstores. *OSHA Safety and Health Standards,* Code of Federal Regulations, Title 29, Part 1910, contains regulations relating to the safety features to be included by the agency or architect-engineer in the design of any project. *Construction Safety and Health Regulations,* Code of Federal Regulations, Part 1926, pertains specifically to construction work.

It was OSHA's intent in preparing Part 1926 to place in one volume all of the rules and regulations applicable specifically to construction work. It is suggested that each of the contractor's superintendents, foremen, or other supervisors have a copy for reference. To the extent that Part 1926 contains standards that are incorporated by reference, copies of applicable referenced material should be made available to the supervisors as well.

In order to ensure compliance with applicable regulations, contractors should assemble copies of the pertinent regulations that are incorporated by reference, as they relate to those regulations applying to the specific project under construction. OSHA documents can be inspected at OSHA in Washington or at any of its regional or field offices.

The Act deals with all working conditions and includes the following broad categories:

OSHA AND THE CONTRACTOR

General safety and health provisions.
Occupational health and environmental controls.
Personal protective and life-saving equipment.
Fire protection and prevention.
Signs, signals, barricades.
Materials handling, storage, use, and disposal.
Tools—hand and power.
Welding and cutting.
Electrical.
Ladders and scaffolding.
Floors and wall openings and stairways.
Cranes, derricks, hoists, elevators, and conveyors.
Motor vehicles, mechanized equipment, and marine operations.
Excavations, trenching, and shoring.
Concrete, concrete forms, and shoring.
Steel erection.
Tunnels and shafts, caissons, cofferdams, and compressed air.
Demolition.
Blasting and use of explosives.
Power transmission and distribution.
Rollover protective structures; overhead protection.
Recording and reporting work injury frequency and severity data and accident cost.

These categories are further divided into sections listing specific requirements, including the posting of certain notices that projects are covered by the law, and submittal of certain data within the time limits on standardized forms.

Safety Plans

When bidding a project, the contractor should take into account how much it will cost to run a safe job. Obviously, if the project involves ultrahazardous activities such as blasting, tunneling, or the interruption of vital public services, the contractor will have to allocate more of its project budget to safety. Even on jobs that are not ultrahazardous, the contractor should budget for the implementation of a safety plan or program. A good safety record takes years to establish, but only moments to destroy.

In the development of a *safety plan* or *safety program,* an estimator familiar with safety requirements should review the project and prepare a budget for a plan to protect the contractor's employees and agents, the project itself, and the public and its property. The contractor's superintendent or project manager should also be involved in the preparation of any budget, as project safety will, on a day-to-day basis, be the superintendent's responsibility.

After the award of a contract, the contractor's superintendent or project manager and any others who will be held responsible for safety on the project should develop and implement a specific safety plan for the project. The plan should then be presented by the project superintendent or project manager or another individual held responsible for safety on the project to the various subcontractors on the project at the time of the preconstruction conference or at the first jobsite meeting. Of particular importance will be the discussions and agreements reached with those subcontractors who will be performing the excavation, sheeting and shoring, foundations, or any other unusually hazardous work on the project.

As the job progresses, safety needs will change and develop, and new concerns such as delivery access, hoisting facilities, and working schedules will become important. It is necessary for the superintendent or project manager to continue the practice of discussing and reaching an agreement with each subcontractor that comes onto the job as to its safety responsibilities on the project. It is a good practice for the superintendent or project manager to make a record of all agreements reached with each subcontractor regarding that subcontractor's safety responsibilities and a dated log of all safety conferences and meetings.

Although specific meetings on safety may not be warranted or economically justified on each project, safety requirements and ongoing compliance with the safety plan should and can be a recurring item on any jobsite meeting agenda. By making safety and compliance with the safety plan an ongoing agenda item, all parties involved in a project will remain more aware of safety as the project progresses. Remember, a safety plan only works when the individuals actually doing the work perform in a safe and workmanlike manner. Any method that can implement the safety plan, such as foreman meetings, a job safety officer, posters, signs, awards, equipment maintenance programs, inspection checklists, and the like, should be employed.

Record Keeping

Although OSHA only requires contractors with 11 or more employees to maintain and make available records of occupational injuries and related illnesses, the record-keeping requirements are good practice even if a contractor has fewer employees. Although not every injury or illness occurring in the workplace is recordable, records should include such items as fatalities (see Figures 6.1 and 6.2), lost work days, and incidents requiring medical treatment. Fatal and serious accidents must be reported within 48 hours.

Records or copies of the records must be kept on prescribed forms and generally three types are involved.

1. Log of injuries or illness with entries to be made within six days of occurrence.
2. Supplementary records setting forth additional details.
3. An annual summary.

The contractor is required to keep such records for a period of five years.

Safety Violation Procedures

The Labor Secretary, or an authorized representative, is authorized to enter and inspect all working places covered by OSHA. Such inspections must take place in a reasonable time, within reasonable limits, and in a reasonable manner—normally during business hours. Usually, inspections are conducted by a team of experts from OSHA, code compliance inspectors, or similarly authorized experts under state plans. Generally, there is no advance notice of an oncoming inspection.

The inspection is divided into three parts, the first of which is the opening conference with the contractor. The inspector will normally present his or her credentials and explain to the contractor the nature and scope of the inspection. An employee representative, as well as a representative from the contractor, are selected to participate in the physical inspection of the workplace.

The next step is the actual physical inspection of the workplace itself, which is called a "walkaround." During the inspection, the inspector will walk through the workplace and examine conditions and note apparent hazards. The inspector may stop and talk to employees at the site regarding safety techniques and practices in the workplace. OSHA specifically calls for

REPORT OF CONTRACTOR'S ACCIDENT

Date: 23 Aug 1988
DAY: T (XX)
WEATHER: Clear (X), Rain (X)
TEMP: 70-86 (X)
WIND: SW (X), No. 1

Project: Arroyo WWTP Addition
Unit:
Proj. No.: 88-0324   Contract No.: 07-01-789224
Contractor: B&H Construction, Inc.
Sub-Contractor: n. a.
Date of Accident: 23 Aug 1988   Time: 9:20 AM   Location: City of Casa Grande
Description of Accident: Site cleanup was in progress after form removal. Laborer was sweeping ledge approx. 30 ft above concrete floor. Rectangular hole which was previously covered with a plywood sheet was uncovered by laborer to sweep dirt through. While sweeping, laborer backed into hole.
Primary Cause: Walking backward while sweeping, worker stepped into hole and fell to his death on concrete slab 30 ft below.

Contractor's Personnel or Equipment

Name of Injured Employee: James L. Martin   Age: 29
Occupation: Laborer   Sex: M
Nature of Injury: Skull fracture
Degree of Injury: First Aid ☐ Doctor Visit ☐ Hospital ☐ Fatality ☒

Type of Equipment: n. a.
Extent of Damage: n. a.

Other Persons or Property

Name of Injured Party: No other persons involved   Age:
Address: City: State:
Nature of Injuries:

Name of Property Owner: City of Casa Grande   Address:
Nature and Extent of Damages: none

Data used are fictitious for illustration only

Was Use or Lack of Safety Equipment a Factor in This Accident: No
If so, Explain: n. a.

What Safety Regulations Were Violated: None. Walkway had been properly protected with guard rails. Floor hole had been properly covered prior to accident.
What Corrective Action Has Been Taken by the Contractor: Cautioned other employees about watching where they are going and using special care working about the site.

DISTRIBUTION:
1. Project Manager
2. Legal Staff
3. Engineer/Architect
4. Project File

Report by: [signature]
Title: Project Manager
LFM Constructors, Inc.

Wiley-Fisk Form 8-13

FIGURE 6.1 Report of contractor's accident.

FIGURE 6.2 Photograph of accident scene accompanying the report in Figure 6.1.

consultation with employees in private, if required. An employee may point out to the inspector conditions that he or she believes are hazardous.

The final stage of the inspection is the closing conference. After the inspector has completed the inspection, the observed conditions and safety hazards are discussed with the contractor's representative. The inspector should then explain what citations may be issued for some or all of the violations observed and that monetary penalties may accompany each citation. If citations are issued, a reasonable length of time will be allowed to correct the situation. If a contractor is fined and believes it to be unwarranted, the contractor has 15 days to appeal any penalties.

Violations under the Act are criminal in nature. Violations resulting in a citation can result in a civil penalty of up to $1000, and even up to $10,000 if the violation is willful or repeated. For failure to correct a violation within the time allowed in a citation, a civil penalty of up to $1000 per day for each day the violation continues may be assessed. For a willful violation that results in the death of an employee, a first conviction can result in up to a $10,000 penalty and a jail term of up to six months, and a second conviction can result in fines of up to $20,000 and as much as a year in jail.

Although an insurance company will not be responsible for violations or penalties arising out of any such violations, it may become involved if an accident is involved. Any failure of the contractor to conform to OSHA regulations could be considered negligence in itself, which would directly involve and have an impact on the insurance company. The contractor's compliance with OSHA regulations is therefore very important to the contractor's insurance company as well as to the contractor itself in terms of obtaining lower insurance premiums and insurance coverage. Should an accident occur on a project, a clear job safety record and a good safety plan can be very helpful to a contractor when it comes to ultimate liability and future insurability.

CONTRACTUAL SAFETY REQUIREMENTS

For a number of years, contracts for the construction of public works projects have often included safety requirements as a contractual requirement as well as a legal obligation. Failure to adhere to these requirements places the contractor in breach of contract as well as in jeopardy with OSHA enforcement officers.

When the contractor's safety obligations are not mentioned in the prime contract documents, compliance is primarily a legal obligation between the contractor and the state or federal agency administering the OSHA provisions. However, when compliance with the OSHA requirements is specified in the prime contract documents, it becomes a contractual obligation in addition to a legal obligation. As a contractual obligation, the agency may consider any failure to comply with safety requirements a breach of contract and a basis to withhold payments. The agency or an architect-engineer responsible for the administration of the contract will be responsible for evaluating the contractor's compliance with safety requirements and will determine or make recommendations for payment accordingly.

ACCIDENT RECORDS

The keeping of accident records is an important part of a construction company's safety and health program. These records serve to pinpoint the locations and underlying causes of job accidents and provide information that is vital to the planning of more effective accident-prevention programs. An example of a contractor's lost time accident report is illustrated in Figure 6.3. Although not a substitute for the specific documents required by OSHA in case of an accident, such reports do provide information on the effectiveness of an overall safety effort and how one company compares with other construction firms.

Accurate, complete, and detailed accident records can be invaluable in defending against charges of safety law violations or claims for damages. An initial, on-the-spot report should be made right after the occurrence of an accident (see Figure 6.1).

Exhibits in the form of photographs taken of the accident site, such as the photograph shown in Figure 6.2 that accompanied the previously illustrated accident report, should support such written reports. A video camera and recorder can be used to make a video record that will include both sight and sound. Apart from their role as tools of accident analysis and prevention, these records must conform to the demands of insurance companies and construction contracts with the legal requirements imposed by local, state, and federal agencies. Prior warnings to subcontractors or employees should be logged as well.

FIGURE 6.3 Contractor's lost time accident report.

THIS SIDE OF REPORT TO BE COMPLETED BY CONTRACTOR'S SAFETY ENGINEER OR SAFETY REPRESENTATIVE.

Type of Accident	1. ☐ Railroads 2. ☐ Water 3. ☐ Elevators 4. ☐ Vehicles 5. ☐ Pressure Equipment 6. ☐ Explosions 7. ☐ Fires 8. ☐ Electricity 9. ☐ Flash Burns	10. ☐ Dust—Chemicals—Gases 11. ☐ Handling Material or Equipment 12. ☐ Falling Objects 13. ☒ Falls of Persons 14. ☐ Jumping To or From Places 15. ☐ Striking Against Material 16. ☐ Flying Particles 17. ☐ Hand Tools 18. ☐ Machinery 19. ☐ Not Otherwise Classified

Data used are fictitious for illustration only

Mechanical or Physical Causes

Check One or More and Explain

☐ Layout or Procedure _____
   (Unsafe Method, Arrangement, etc.)

☐ Work Area _____
   (Poor Housekeeping, Lighting, Ventilation, Loose Rock, etc.)

☐ Tools and Equipment _____
   (Improper, Defective, Unguarded, etc.)

☒ Other Unsafe working practice by employee. Employee error

Personal Causes

Injured Person	Other Person	
☒	☐	Unsafe Act Failure to take proper care and attention
☐	☐	Physical or Mental Defect _____
☐	☐	Lack of Knowledge or Skill _____
☐	☐	Other _____

Supervisory Fault

Was Inadequate or Faulty Supervision or Foremanship a Cause or Contributing Cause of Accident?

No
(Explain)

Corrective Action

What Has Been Done to Prevent Similar Accidents in the Future?

Caution all employees of the hazards at a construction site, and remind them to exercise due care at all times

_____   _____
Contractor Project Manager   Owner/A&E Project Manager

Wiley-Fisk Form 8-14b

FIGURE 6.3 (*Continued*)

CHAPTER 6 CONSTRUCTION SAFETY

PROTECTION OF THE PUBLIC

One very important aspect of a contractor's safety program is the safety of the general public. People are naturally curious and capable of many thoughtless actions in their attempts to see what is happening at a jobsite. The contractor must resign itself to the fact that the public wants to know what is happening; it is up to the contractor to determine how they find out. If an attempt is made to shut out people completely, many will feel compelled to climb over fences, follow trucks through the gates, or act in some other equally hazardous way to achieve their ends. Warnings are of little good. One of the best methods to solve this problem of security is to allow the public to view proceedings from a controlled vantage point. The contractor who provides means for the public to see the work and at the same time be protected from the hazards of construction is making a wise move.

The concept of controlled vantage points works well on the construction of public buildings. However, on street work, curb-and-gutter jobs, storm drain construction, sidewalks, pipelines in city streets, and similar public works construction, new hazards present themselves. The contractor may find that the hazard of open trenches in the vicinity of a school, for example, requires that the excavation be protected by chain-link fences during all times while construction is not actually underway. Excavations in busy streets, instead of allowing barricades to be erected at night, often require the contractor to cover all open trenches with heavy steel plates adequate to support traffic loads. Work on city streets often presents other safety problems as well.

Many times, traffic control must be maintained to allow maximum utilization of the roadway during peak traffic periods. This may involve detours, flaggers, special traffic control devices, or other means. The usual standards for such controls are contained in a Federal Highway Administration publication entitled *Work Zone Traffic Control, Standards and Guidelines,* (ANSI D6.1) which is a part of a publication entitled *Manual on Uniform Traffic Control Devices.*

For public works construction, traffic requirements will not only be a part of the contractor's legal obligations, but are often contractual as well. In many cases, traffic requirements are spelled out in detail in the project specifications. Whether or not they are in the specifications, a contractor will be bound by the terms and conditions of a street or police department permit that the contractor was required to obtain, the terms of which the contractor never sees until after being awarded a contract.

TRENCH AND EXCAVATION SAFETY

If a contractor plans to cut corners, the failure to provide adequate trench or other excavation safety is *not* the way. As a result of research conducted by the Associated General Contractors of America and reported in the American Society of Civil Engineers *Journal of the Construction Division,* some revealing facts were discovered regarding excavation cave-ins.

1. At least 100 fatalities occur each year from cave-ins. That is at least 11 times as many fatal accidents as accidents involving disabling injuries.
2. The majority of these cave-ins occur in shallow excavations, primarily sewer trenches.
3. No part of the country is immune to cave-ins.
4. Every type of soil is susceptible to cave-ins.
5. Most cave-ins occur in unsupported excavations.

FIGURE 6.4 Trench safety hazard: Spoil bank too close to edge of trench, plus added surcharge load of pipe on top of spoil bank.

FIGURE 6.5 Example of a properly shored trench for an 87-in. diameter force main.

6. Major factors influencing trench failure are the presence of construction equipment or materials near the edge of an excavation and adverse climatic conditions. Figure 6.4 shows a spoil bank too close to the edge of the trench, with the added risk of a surcharge load in the form of a pipe lying on top of the spoil bank.

7. Usually, engineers do not specify shoring requirements prior to bidding. (In California public works contracts, however, the Labor Code requires the agency to show "Sheeting, Shoring, and Bracing, or Equivalent Method" as a separate pay line item on all contracts involving trenches five feet deep or deeper.)

8. According to the AGC research, approximately 50 percent of the contractors surveyed would prefer the engineer to specify the shoring requirements prior to bidding.

It is quite probable that all of the excavation cave-ins of record could have been prevented if there had been a properly designed supporting system (see Figure 6.5). The added construction costs could well have been compensated for by a reduction in the liability costs.

Trench and excavation shoring is one of the critical safety hazards referred to in the OSHA Construction Safety and Health Regulations. Numerous fatalities have resulted from the failure of a contractor to provide adequately for worker safety under these conditions.

Details of safety codes vary from state to state, but there is a trend toward greater uniformity, and the safety codes of each jurisdiction should be carefully checked prior to beginning work in another state, to confirm the specific limitations and regulations that will control the work.

CHAPTER 7

Bidding Phase

- Selection of Subcontractors
- Prebid Inspections and Conferences
- Disclaimers in Bid Solicitation
- Preparation and Submittal of Bids
- Effects of Weather on Bid Price
- Negotiation of Bid Prices
- Product Substitutions
- Owner Disclosure of Site Information
- Schedule of Values

CHAPTER 7 BIDDING PHASE

The success or failure of a contractor's venture into the world of public works projects hinges on the estimating of a job, the preparation of the bid, recognition of the fundamental differences in actually constructing a public project, subcontract limitations, and the bid submittal.

The bidding phase is critical to the success of a contract, and in order to avoid the pitfalls that so often befall the unwary or unsuspecting contractor bidding its first public project, it is important to at least recognize the numerous issues that must be considered in the preparation of a bid, such as critical inspections or potentially complex administration procedures. The basic rule, however, is that a contractor can and should only bid what is in the contract documents.

Although the initial reaction of a contractor on its first public project to all the apparent meaningless interference with its work progress and ability to get the job done is one of frustration and disgust, such requirements are normal on public projects. The added costs of working under these conditions and the delays resulting from such critical inspection and complex contract administration procedures must simply be calculated into the bid price for a job. Failure to do so will result in a nonprofitable job. It is important to learn the rules of the game and then play by those rules. There are many ways of getting profitable extras from a public client by using the system versus fighting it.

To begin with, the Contract Documents must be read carefully, and it should be presumed that every provision will be strictly enforced. Remember

Education is what you get when you read the contract.

Experience is what you get when you don't read the contract.

It is foolhardy to pass over the General Conditions of the Contract lightly as though they will have no significant effect on the construction operations. Public contracts are viewed as a contractual relationship between the contractor alone and the public collectively. There is a strong tendency for the courts to lean in favor of the public. So remember, the Contract Documents were prepared for a reason, and that reason was probably not to help you. A contractor should factor into a bid the extra cost of compliance with all General Conditions work as well as the added cost of administering a public contract. It is not valid to assume that a contractor's primary responsibility is to merely construct a quality project, without regard for the importance of the administrative procedures involved. Public contracts do not lend them-

SELECTION OF SUBCONTRACTORS

selves well to short cuts in terms of either the procedures or the construction materials and methods.

SELECTION OF SUBCONTRACTORS

After bid opening, *if awarded the job,* contractors have historically started shopping for a different specialty contractor that is willing to cut the prices of the proposed subcontractors whose bids were used in the general contractor's bid. On public works projects, the practice may be found to be illegal and subject to cash penalties in some states. This practice, called bid shopping or bid peddling, has all but been eliminated in some localities by the successful lobbying efforts of subcontractor associations. Bid shopping unfairly puts the subcontractor through the high cost of preparing a bid for a contract that may never be awarded; while the bid may only be utilized by the general contractor as a lever to solicit cut-rate bids from other subcontractors. This practice should not be taken lightly even if not illegal per se in a particular locality, as there is case law in support of subcontractors that, having felt they were being cheated out of their due, have sued in court and won damages to cover lost profits. All is not lost, however, as the situation is never hopeless until the contract is signed. By careful manipulation of cost-influencing items *just before the bids are opened,* an improvement in the profitability of a project can often be realized (see Figure 7.1).

Generally, in those states not prohibiting bid shopping, if a general contractor relies on a trade contractor's firm quotation without making any promises or representations, the general contractor will be in a position to

FIGURE 7.1 By careful manipulation of cost-influencing items just before the bids are opened, an improvement in the profitability of a project can often be realized.

• 135 •

either hold the subcontractor to its price or to elect to do business with another subcontractor. However, in the case of *Electrical Construction & Maintenance Co., Inc. (ECM), v. Maeda Pacific Corp* [764 F.2d 619(9th Cir. 1985)], the U.S. Court of Appeals, Ninth Circuit, found that the general contractor had promised to award the subcontract to ECM if it was the lowest electrical bidder and if Maeda itself was awarded the contract; Maeda, the general contractor, was therefore under a legal obligation to award the subcontract to ECM.

Another obligation frequently imposed on a general contractor bidding a public project is the requirement that a certain percentage of the work be done by the contractor's own work force. This is essentially an antibrokerage requirement designed to eliminate the bidder that plans to do none of the work with its own forces, but only to provide the construction management of the work of the various subcontractors. The amount of work thus required of the general contractor often varies from not less than 15 percent of the total value of the contract to as much as 50 percent. In a Board of Contract Appeals decision, [*Appeal of R.M. Crum Construction Co.,* VABCA No. 2158 (May 10, 1985)] the contractor had its contract price reduced due to a failure to perform a sufficient percentage of the work with its own forces. However, in that case, the terms of the contract stated that if the contractor failed to meet the requirement, the contract price was to be reduced by 15 percent of the value of that portion of the percentage requirement that was performed by others. Without such a provision in the contract, the enforcement of a penalty is highly in question. In the case just mentioned, the contractor had based part of its percentage on the value of a large amount of materials supplied by the contractor; however, the materials were being installed by its subcontractors and thus did not qualify as part of the general contractor's construction value.

PREBID INSPECTIONS AND CONFERENCES

Whenever a public agency provides for a prebid inspection or tour, or a prebid conference, all general contractor bidders should plan to attend. Often, significant facts can be ascertained from such inspections, tours, or conferences that may seriously affect a bidder's ability to construct the work in accordance with practices that may be considered customary on other jobs. Such inspections, tours, or conferences may be limited by the public agency to a specific time only, and no assumption should be made that a site will be made available for inspection at the bidder's convenience. This was aptly demonstrated in one matter involving a contractor that attempted to inspect a project site a few weeks after the designated date. When the government denied access to the contractor, a protest was filed, claiming that such actions were unreasonable. In a decision of the Comptroller General, [BECO Corporation BCA File No. B-217573 (May 15, 1985)] it was ruled that the government was entitled to limit prebid site inspections to a single date. In denying the validity of the contractor's protest, the Comptroller General said that the government is under no obligation to accommodate the preference schedule of each individual bidder.

An agency may limit site visitations to a single time period in order to assure that all prospective bidders receive the same information. This is particularly true of projects involving the construction of an addition to an operating facility such as a treatment plant, where unscheduled visits can create both the disruption of plant operation and, in some cases, a security risk.

DISCLAIMERS IN BID SOLICITATION

Frequently, in public contracts, disclaimers will be included to place all of the risk of unknown conditions on the bidder. The effectiveness of such provisions in the contract appears to be highly in question, but the risk of loss to the bidder is still present until the final resolution of each specific case. One example is the case decided in a Georgia court in which the court enforced a public owner's disclaimer of the adequacy of designated borrow pits and disclaimer of the accuracy of the accompanying boring logs. [*Jahncke Service, Inc. v. Department of Transportation*, 322 S.E.2d 505 (Ga. App. 1984)] In that case, the Department of Transportation disclaimed responsibility for borrow materials by stating in the contract documents

> The Department, in making this borrow report available to Contractors, assumes no responsibility if the contractor relies on this information. . . . The quantity of material shown on the plans as available in the borrow areas is not guaranteed. . . . The obligation is upon the Contractor, before making his Bid or Proposal, to make his own investigation.

During construction, it was discovered that the designated borrow pits did not contain the quantity or quality of materials anticipated from the logs of borings and accompanying report, and the contractor was forced to obtain materials from another source at some added cost.

The Department of Transportation relied on the disclaimers for their defense, and the Georgia Court of Appeals agreed that as a matter of law, the contract provisions precluded recovery by the contractor.

It should be noted, however, that some other courts have held that disclaimers such as the one just mentioned will not be enforced in a public contract. They base their position on the fact that the short period of time allowed for bid preparation makes it impossible to conduct an independent subsurface investigation and forces the contractor to rely on the agency's representations, regardless of the disclaimer. It should be kept in mind that the bidder is generally not obligated to seek expert engineering and geological advice and interpretation of subsurface data prior to bid submittal. [*Appeal of Blake Construction Co. and U.S. Industries,* a Joint Venture ASBCA No. 20747 (March 25, 1983)]

Where on-site drill holes are available for inspection by the contractor, it is wise for a bidder to avail itself of this opportunity to measure the water levels in the holes, as the resulting information has the potential of advising the bidder on the differences in ground water conditions since the time of the original borings. It is never reliable to base a bid upon the ground water conditions indicated in the original logs of drill holes, for such information represents conditions that may have existed at another year or time of year and are thus inconclusive, and frequently differ from present conditions.

Notwithstanding a proper disclaimer of subsurface information, the agency providing the subsurface information still has a duty of nonnegligence in the preparation of the drill logs and site information.

PREPARATION AND SUBMITTAL OF BIDS

Qualifying of Bids

A popular notion in private contracting is the time-established practice of qualifying a bid. That is, altering some of the provisions on the bid sheet, or inserting provisos that limit the bidder's responsibilities to the conditions inserted by the bidder. In public works contracting, this is grounds for the rejection of a bid. No qualifying statements of any kind or alterations to the original offering should be made on the bid submitted to a public agency in response to a publicly advertised project. A public contract is referred to as a contract of adhesion. This means, in short, that all of the terms and conditions of the contract are prepared by the agency and are not open to alteration or negotiation of any kind by the contractor: In short, it is a take-it-or-leave-it contract. It is often said that "a bid opening is a poker game in which the losing hand wins."

A bidder who pays the cost of preparing an estimate and submitting a bid is foolish to believe that by making minor modifications to the bidding documents it can obtain a tactical advantage over other bidders. What actually may happen is that the public agency must reject such a bid, and all of the bidder's preparation costs are lost.

Responsive Bids

Often, you may have heard the term *"lowest responsive, responsible bidder."* The award of public contracts must be made to the lowest responsible dollar bidder that is responsive. The term *responsive* simply means that the bidder has complied strictly with each and every requirement stated in both the notice inviting bids and the instructions to bidders. Failure to do so will render a bid *informal* or *nonresponsive* and is grounds for rejection.

In one case, the authors recall that the low bidder on a public project failed to sign some innocuous document but offered to do so at the bid opening. The second low bidder objected and threatened a lawsuit if the award was made to the low bidder because, in the words of the second low bidder, the low bid was nonresponsive because of its failure to follow all instructions exactly. The low bidder responded by stating that the oversight did not affect the competitive nature of the bid and that it was a minor discrepancy and was correctable. The city attorney recognized that the outcome of such a trial was quite predictable, but nevertheless the project would be delayed while the issue was being resolved. Such delays could shift the project into a later time period that would increase the likelihood of delay claims due to weather problems, escalation, impact costs, or other problems. Thus, the city attorney felt that the only course open to the city was to reject all bids and readvertise the job.

To show the degree of concern over apparently minor informalities in preparing a bid and to emphasize the importance of strict compliance with the instructions to bidders, the following case is cited: Ameron, Inc. submitted a low bid on a federal project and attached the Standard Form 24 (federal) Bid Bond form. The bond stated that the penal sum was 20 percent of the bid price, not to exceed a typewritten amount of $3,000,000. The numerals had been typed over a "whited out" area in the box provided in Standard Form 24. The bid was thereby rejected by the government as being nonresponsive. Although affidavits later showed that the change had been made by the typist for the bonding company, altered bonds are unacceptable unless accompanied by evidence that the surety consented to the alteration. The comptroller general upheld the rejection of the bid, stating "the submission of a materially altered bond can have the same effect as the failure to submit a bond altogether, because under surety law no one incurs a liability to pay a debt or to perform for another unless expressly agreed to in the bond. An alteration in the bond thus raises a question whether the surety agreed to the altered terms." [*Ameron, Inc.,* BCA File No. B-218262 (April 29, 1985)]

There is case law in California in support of the concept that even an unsigned bid may be considered as valid; however, it would require a court to decide the issue and meanwhile the job would be delayed. Job delays affect a contractor's costs, of course, and a bid price that was submitted three months ago may not be considered as attractive by the bidder as it did at the time of submittal.

Every document that is requested by the public agency must be carefully filled out and submitted. Often, where there are federal or state funds involved, as in many grant programs, additional forms and documentation are required. These are not window dressing or a matter of just going through the motions. The issues addressed in these documents are rigidly enforced, so the costs of compliance should be included in the contractor's bid.

Unit Price vs. Lump Sum Bids

Although generally uncommon in private construction contracts, the concept of unit price bidding is frequently used in civil engineering construction on public works projects. It is a system that does not lend itself well to the construction of buildings, but for roads, curbs and gutters, storm drains, sanitary sewers, waterlines, canals, dams, earth embankments, flood control, irrigation, concrete structures, and similar work, it works quite well. However, it must be remembered that the cost of administering a unit price contract is likely to exceed that of administering the same contract under a lump sum price, partly due to the higher cost of the methods necessary for verifying work quantities for payment.

The principal difference lies in the fact that under a lump sum contract, the bottom line price is the bid, and that as long as there is no valid differing site conditions claim, or as long as the price has not been amended through the issuance of change orders, any cost overruns whether or not they are the result of quantity overruns are the contractor's risk. Personally, the authors favor lump sum contracts because they allow maximum flexibility for manipulating cost-influencing items of work, and with competent management they can be one of the most profitable methods of contracting if a competent set of plans and specifications is available and design conditions are fixed. The risks are high, but so is the profit potential.

In a unit-price contract, each item of work is listed separately on the bid sheet along with the engineer's estimate of the quantity of that item. The bidder then inserts a unit price, multiplies the unit price by the engineer's estimated quantity of that item, and inserts the product in a column provided for extensions. To determine the low bidder in order to make the award, the column of extensions is added up and the bottom line total becomes the bid price solely for the *purpose of comparing bids.* However, the principal difference between lump sum and unit price contracts lies in the fact that the bottom line is *not* the final bid price. But rather, each *unit price* represents the contractor's bid. Thus, if the estimated quantities as determined by the engineer differ from the actual quantities measured on the job, a new extension is computed using the original unit price multiplied by the new (actual) quantity. The total price that the contractor receives for the job is the sum of all of the new extensions plus the value of any change orders.

Herein occasionally lies another problem. It is customary for the public agency to provide in the terms of the contract that the contractor must guarantee that the unit price will apply to all quantity variations up or down from the quantities shown in the engineer's estimate, thus causing a monetary loss every time a quantity change is made that results in a separate shipment or product returns. To partially mitigate this situation, the guarantee is limited to a percentage of the engineer's estimate. The Associated General Contractors of America (AGC) suggests a guarantee of 15 percent, and the government under the Federal Acquisition Regulations uses 15 percent; however, it is not uncommon to find the guarantee at 25 percent in contracts with other public agencies. This percentage limits the owner's rights to demand that excessive quantity variations be furnished at the originally quoted unit cost and gives the contractor a bargaining position.

An example of its use would be a case in which a job called for 10 valves at, say, $1000 per valve. Subsequent to the ordering and delivery of the 10 valves, let us say that the engineer determined that 11 were needed, and the contractor now had to order one extra valve. As the quantity represents a quantity change of less than the 15 or 25 percent guarantee, the contractor is contractually obligated to furnish the extra valve for $1000, even though separate shipping costs, handling costs, and the lack of a quantity discount may create a financial hardship for the contractor, such that to make the same profit that was estimated for the original 10 valves, the contractor would normally ask $1200 for the same valve, but must sell to the owner for $1000. Conversely, if the engineer had decided that only nine valves were needed, the contractor would be required to credit the owner with $1000 for the returned valve, in spite of the fact that it would involve return charges, shipping and handling costs, time lost, and lost profit. In addition, frequently a contractor will apply corporate overhead and profit as a straight line across the board so that each pay line item carries a share of this overhead and profit, even though it has nothing directly to do with the individual pay line item (a risky procedure). Thus, a fair credit to the agency might actually be only $800 for the unneeded valve instead of the $1000 that you are obligated to return under the terms of the contract, because the deletion of the one valve does not exceed the 15 to 25 percent allowable quantity variation specified.

The contractor should watch quantities closely. As soon as the quantity variation exceeds the allowable guaranteed under the contract, the contractor is in a position to increase the prices on quantity increases or to decrease the unit prices on quantity decreases so as not to sustain the more severe losses that would otherwise occur. This is money that the contractor has earned, and it should have every right to make such price adjustments when the quantities vary excessively. The quantity variation clauses are an equitable means of resolving a problem that could otherwise hurt either the owner or the contractor.

Time of Submittal of Bids

Typically, in private construction, bids may be solicited privately from selected contractors formally or informally, and the parties may then negotiate the terms of a contract. In public works contracts, all projects must be advertised publicly according to specific laws of the state in which the work is to be performed. Furthermore, a specific time and place must be set for public opening of the bids, and the time of opening bids must be rigidly adhered to. If the solicitation for bids states that the bids will be opened at 10:00 A.M., the public agency is prohibited by law from accepting any bid submitted after that time. A bid that is submitted even a minute or two after the specified time of opening requires disqualification and the bid must be returned unopened.

Sometimes, bids are submitted by mail and fail to reach their destination in time. A case is documented in which a bid was submitted by mail on time but it arrived late, even though it was postmarked five days before the specified date of bid opening. When the agency attempted to disqualify the bidder for having been late, the matter was taken to court and the court judgment was that the contract had to be awarded to the late bidder (who was low, of course) because its bid had been postmarked five days before the deadline. As stated by the court, a bid's being postmarked five days before the bid opening day was acceptable as evidence of a timely submittal, even if its arrival was delayed by the mails. However, in the case of *Rainbow Roofing, Inc.,* [BCA File No. B-213515 (June 27, 1984)] the Comptroller General ruled that a postal meter impression was not acceptable as evidence of the date of mailing, and that in order for a late bid to be accepted, a hand-cancellation bullseye is required on both the envelope and the certified mail receipt. Meter impressions, whether imprinted by a postal employee or a private party, are not acceptable. On federal projects, this is covered under 48CFR14.304-1(a)(1) of the Federal Acquisition Regulations.

Bid Bonds

Although in private contracts, bonds are sometimes required, under a public contract, certain bonds are always mandatory. Along with the bid, the bidder is required to post a bid security either in the form of a cashier's check or a bid bond in the sum of five or ten percent of the total bid price for completion of the work. (For Federal Projects see 48CFR28.101-2.) In addition, bonds must be provided to cover performance and payment of subcontractors and suppliers. For public projects in most states and for the federal government, a separate performance bond and a separate payment bond (sometimes called a labor and materials bond) are required. (See 48CFR28.102 of the Federal Acquisition Regulations.) In some states, a combination payment and performance bond may be required instead of two separate bonds.

If the total contract amount is adjusted as by change orders, the bond amounts must be increased to cover the added cost. These bonds are normally required to be submitted before the agency will sign the contract.

It should be noted, also, that most public contracts contain a requirement that bonds be issued by a surety approved by the agency. The federal government publishes a list of acceptable sureties for federal contracts in Treasury Department Circular 570. Thus, if for any reason the public agency objects to the surety used in a bidder's proposal, it retains the right to require the bidder to deal with another surety company on its work. A contractor, of course, has similar rights and it may object to a surety proposed by its subcontractors. [See *Turner Construction Co. v. Seaboard Surety Co.,* 469 N.Y.S.2d 725 (N.Y.A.D. 1983)]

Bid Errors—Relief of Bidders

Definition of a Low Bidder: "A contractor who is wondering what he left out." What happens, then, if a bidder does make a mistake in its bid? Is a bidder trapped into a contract that it cannot get out of? Generally, the answer depends upon what kind of an error was made. Of course, a bidder can always default and refuse to sign a contract, but this will result in forfeiture of the bid security and a bad record with the bonding company. Most states pattern their contract administration procedures after the Federal Acquisition Regulations (48CFR14.406); thus, a fair solution is provided.

Bid errors can be divided into two categories: errors of fact and errors of judgment. An error of fact is generally excusable; an error of judgment is not. [Appeal of *Bromley Contracting Co., Inc.,* HUDBCA No. 81-624-C30 (April 18, 1983)] In order to exercise the right to withdraw a bid, a contractor must satisfactorily establish the following:

1 A mistake was made.
2 Proper written notice was given to the public agency.
3 The notice must specify how the mistake occurred.
4 The mistake made the bid materially different than intended.
5 The mistake was made in filling out the bid and was not due to an error in judgment or carelessness in inspecting the site of the work or in reading the plans and specifications.

A further condition is usually that the bidder will not be permitted to correct the error and resubmit a bid on the same project. [*Edmonds Electric Co.,* BCA File No. B-214063 (June 11, 1984); *Mid-South Electric Co., Inc.,* BCA File No. B-213894 (June 14, 1984)] [cf. CA Public Contracts Code §5105] Furthermore, a bidder who was permitted to withdraw its bid because of an excusable error should be permitted to recover its bid security. In some areas, these conditions are express provisions of law. In others, they are merely an exercise of public policy. For federal contracts, the procedures for a contractor to cancel or withdraw its bid are spelled out in the Federal Acquisition Regulations (48CFR14.406).

Opening of Bids

Although in the administration of private contracts, in which informality may even be preferred or at least accepted, no rigid procedures are necessary. On public works contracts, however, the opposite is true. There are a number of inflexible rules for the opening, acceptance, and documentation of all bids received.

Several primary matters of concern during this period include

1. Receipt of sealed bids at the designated time and place.
2. Refusal to receive any bids even a few minutes late. If the Notice Inviting Bids says 2:00 P.M., then at exactly 2:00 P.M. no more bids will be accepted. . . . so, get out of that phone booth and into the hearing room on time.
3. Confirmation that all bids are responsive.
4. Acceptance and logging of each bidder's name and the amount for all responsive bids.

EFFECTS OF WEATHER ON BID PRICE

Often, the subject of disputes between public agencies and contractors new to the administration of public contracts is the customary method of relating to weather-caused delays.

When a contract specifies that normal climatic conditions are to be considered when bidding, disputes will revolve around what constitutes "normal" weather. When a contract is silent on the matter, the contractor must use its judgment. If a project will take more than one year to complete, generally anticipated weather conditions should be considered in the bid. For shorter-term projects, however, it may not be possible nor advisable to take weather conditions into account. A project involving very severe weather conditions would, of course, be an exception.

A contractor's bid must take into account the probable lost time resulting from predictable severe weather conditions without expecting any relief in the form of extensions of time for such delays. In a typical case, a contractor's claim for a time extension because of heavy rain that made it impossible to work was rejected by the Armed Services Board of Contract Appeals, who ruled that weather delays are excusable only when the weather is abnormal, stating "No matter how severe or destructive, if the weather is not unusual for the particular time and place, or if the contractor should have reasonably anticipated it, the contractor is not entitled to relief." [*Appeal of B.D. Click Co.,* Inc., ASBCA No. 24586 (July 27, 1984)] Such weather delays are part of the risk of bidding and should be covered in the contractor's bid.

If, on the other hand, a contractor was forced into bad weather construction due to an owner-caused delay, the owner becomes responsible for damages resulting from the delays. In a case where the work would have been completed prior to bad weather were it not for a government-caused delay, the contractor was compensated for 22 days of weather delay. In that particular case, Unitranco held a prime contract from the Federal Aviation Administration for work at a Louisiana airport. The contract stated that the government was responsible for establishing a baseline and benchmarks at the work site. The resident engineer's inefficiency in performing his task delayed the contractor and caused it to perform work out of sequence. [*Appeal of Unitranco,* DOTCAB No. 1026 (June 30, 1982)]

This delay, plus the resident engineer's insistence that certain materials be prepared in a manner not required by the contract, caused the contractor to miss good working days and pushed the work into a period of adverse weather. An issue then arose as to whether or not the government must compensate the contractor for damages incurred during the weather delays.

The government argued that under the terms of the contract, weather delays entitle the contractor to a time extension, but no price adjustment. The Department of Transportation Contract Appeals Board said that this was true, except when the weather delays could have been avoided but for the action of the government.

NEGOTIATION OF BID PRICES

In private contracts, it is possible to negotiate at any time prior to the signing of an agreement to construct the work. In public works contracts, however, it is against public policy to negotiate a construction contract except in emergency conditions. Bids must be prepared and submitted in sealed envelopes and read publicly at the appointed hour, and the award must be made to the lowest bidder. Upon the selection of the lowest bidder, no discussion can be held about the terms of the contract, as under a contract of adhesion the contractor's only choices are to sign or not to sign. If any adjustment is to be made in the terms of an agreement, they must be made by a change order after the agreement has been signed.

PRODUCT SUBSTITUTIONS

The temptation to base a bid on substitute products for those named in the specifications is great but should not be considered lightly. There are certain risks connected with this practice, and although it may well spell the difference between getting the job or not getting it, the bidder must recognize the risks and be prepared to deal with them if the risks materialize.

The general rule in public agency contracts is that wherever brand name products are specified, the specifier must name one or two such products and the words "or equal." The first rule to remember is that the opinion on what is "equal" is never left to the contractor. The decision, according to all specifications the authors have known, is left up to the architect-engineer whose decision "shall be final." That is, it will be final unless you feel strongly enough about the issue to take it to court; then, the decision will be up to the judge. However, that method is often more costly than supplying one of the brand names specified, so consider the consequences before spending money on lawyers. So what then does the term *or equal* mean? According to the courts, it is defined as equal quality and utility only. Esthetics cannot be a deciding factor, and the architect-engineer cannot insist upon a substitute product that is identical, just one that will perform the job it was meant to do and is of comparable quality to the product specified.

The risk, however, is simply this. If a contractor bases its bid upon one of the named brands, it may be assured that the product must be accepted by the architect-engineer. If, on the other hand, its bid is based upon a product that the contractor considers equal, but which was not originally named in the specifications, the burden of proof as to equality rests solely upon the shoulders of the contractor. The contractor must submit substantiating data to support its claim of equality, submit to the architect-engineer for review and approval, and risk disapproval followed by the selection of another alternative product submitted in accordance with the same procedure. Delays resulting from this process are neither excusable nor compensable as long as the architect-engineer processes the contractor's submittals in the amount of time allowed for review and return to the contractor as stated in the contract documents.

Often, the data that are required to be submitted are not obtainable from catalogs or published information available on the product, but may actually involve computation sheets by an engineer engaged by the contractor or the product manufacturer. Check in advance. If a product manufacturer will not back up a contractor with engineering support, do not risk pushing their product as it will only cost the contractor more money and may even delay the job.

OWNER DISCLOSURE OF SITE INFORMATION

It is a well-established rule of law that a public agency is under an obligation to provide the bidder with full disclosure of all known site information. That is not to say that each bidder is necessarily provided with its own copy of all such data, but all such information must be made available to each bidder prior to the submittal of their bids. Often, soils reports are stated as "available"; then, when a bidder goes to the owner's offices to review them, it finds that no one knows where they are. In such cases, a contractor should be sure to accurately document this event, as it may provide considerable support in a later claim that the contractor may find it necessary to file.

An example may be seen in the case of a federal contract in which the government made it the bidder's responsibility to inspect the site involved, but did not inform the bidder of an unusual condition that could not have been detected during a prebid inspection. On this General Services Administration project, the contractor stripped an existing coat of paint from interior woodwork and discovered an existing prime coat of aluminum paint. The primer could not be stripped chemically and had to be burned off. In a claim of differing site conditions, the contractor argued that the existence of the aluminum paint was highly unusual and not indicated in the specifications. The government argued that it had stripped a two-inch area revealing the aluminum primer. Thus, argued the government, it should have been detected during the prebid inspection.

The General Services Administration, Board of Contract Appeals, ruled in favor of the contractor, stating that although the solicitation made it the bidder's responsibility to inspect the site, a reasonable inspection would not require a bidder to notice a two-inch scraping. The government had superior knowledge of the unusual character of the primer coat and failed to include this information in the specifications. A reasonable prebid inspection could not be expected to reveal this latent condition; therefore, the contractor was entitled to additional compensation under the Differing Site Conditions clause.

SCHEDULE OF VALUES

In lump sum bids, the contractor traditionally only had to submit the bottom line price. There is an increasing tendency both in the United States and other countries to require that lump sum bids be accompanied by a bid breakdown. In some cases, this is referred to as an "Allocation of Contract Price," but more commonly it is known internationally as a "Schedule of Values."

Although there is little question of the right of the owner to require the submittal of the Schedule of Values prior to administering the first payment request by the contractor, its purpose should be emphasized. The purpose is solely for computing the value of the work done each pay period so as to arrive at a mutually agreeable progress payment amount. It does not contractually obligate the contractor to the unit prices named in the Schedule of Values for change order pricing or the value of extra work. Superficially, a Schedule of Values will resemble the unit price bid, but keep in mind that it is created by the contractor solely to facilitate monthly payments, so it should be balanced accordingly. Measurement of quantities for lump sum projects will be based upon approximations of the quantities of the various items of work accomplished each pay period, but in the end, the amount received by the contractor must be the amount of the lump sum bid as modified by contract changes. Thus, there is no protection for either the owner or the contractor from unanticipated quantity overruns or underruns, unless they are covered under a differing site conditions clause or a change order.

Preconstruction Phase

CHAPTER · 8 ·

- Award and Contract
- Insurance; Bonds; Permits
- Subcontracting
- Preconstruction Conference
- Notice to Proceed
- Schedule of Values on Lump Sum Jobs

CHAPTER 8   PRECONSTRUCTION PHASE

AWARD AND CONTRACT

There are four primary steps involved in the preconstruction phase (see Figure 8.1), beginning with the date of opening bids.

Bid opening date.
Date of award of contract.
Execution or signing of the agreement.
Notice to proceed.

On public works contracts, the procedures are formalized and, in most cases, agencies will follow all four steps. However, some may skip step four, with the result that the starting date of the project may not be known or confirmed by the contractor until several days to a week after the date that the site was available to start construction. This occasionally gives rise to contractor claims for lost time and requests for an extension.

INSURANCE; BONDS; PERMITS

Bonding and insurance requirements are always spelled out in the Conditions of the Contract. The scope and limits of coverage of all insurance policies are specified and are not a negotiable issue. Performance and payment bonds are required on all public works contracts, and the amounts vary, but the most common requirement is for each to be in the amount of 100 percent of the total contract price. Note the words *contract price,* not *bid price.* This effectively means that if the amount of a contract price is increased through a change-order process, the contractor is required to increase the bond coverage proportionately. Thus, a contractor should be

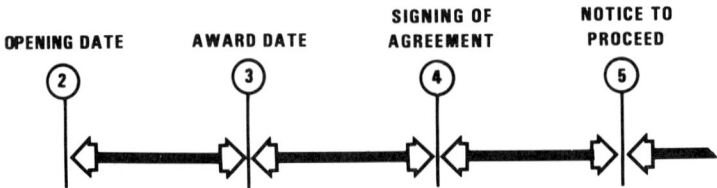

FIGURE 8.1   Award and contract procedure.

careful to factor this into the price of any change order. In some states, instead of two separate bonds, a single combination payment and performance bond is required.

On public works contracts, all bonds and insurance usually must be submitted no later than the time of signing the agreement (point 2 in Figure 8.2). The latest time for obtaining and submitting bonds and insurance is after award of a contract, but before the agency signs the agreement. Failure to provide bonds and insurance by this time may result in a contractor's being declared in default and may cause forfeiture of the bid bond, with the award then going to the second lowest bidder. Instead of default claims, the bid may be declared as *nonresponsive.* In either case, you lose the job. For a proposal to be considered responsive, many public agencies require that the required bonds and insurance be submitted with the bid.

Permit requirements should be examined carefully in the contract documents when bidding a job. There are several ways of handling them in a contract. Often, the contractor is obligated to obtain and pay for all permits. Sometimes, when the contracting agency is a regulatory body, it may consent to issue its own permits to the contractor at no cost. The biggest risk is not knowing the scope of responsibilities included in the terms of the permits. Some permits are more than administrative permission to proceed. Permits such as street department or police traffic permits may include costly tasks as conditions of the permit. Many such risks can be shared with the subcontractors, however (see the section on subcontracting in this chapter).

As an aid to the contractor in meeting the various insurance requirements of a contract, an insurance checklist based upon data developed by the

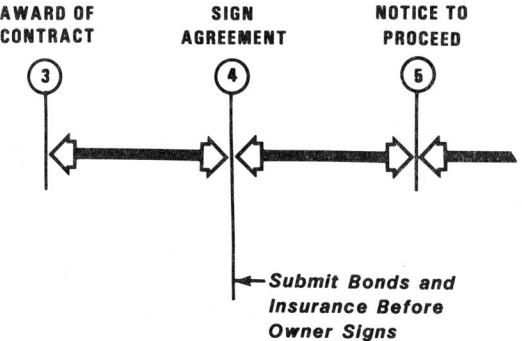

FIGURE 8.2 Submittal of insurance and bonds.

Associated General Contractors of America, the American Insurance Association, and the American Mutual Insurance Alliance, and a bond checklist, based upon data developed by the Associated General Contractors of America and the Surety Association of America, are provided in Appendix A.

SUBCONTRACTING

Unless a job is very small, or you as a contractor are large enough to provide every type of contracting specialty required by the contract, one or more subcontracts will have to be entered into in order to complete the project. As the construction industry has become more complex and contractors more specialized, the amount of subcontracting has correspondingly increased. In some cases, contractors broker a job by subcontracting all of the work and provide only construction management services, financial support, bonding capacity, and overall scheduling and coordination.

Consequently, it is essential that a contractor carefully consider the terms of any subcontract entered with a subcontractor in order to insure that its terms complement and work in conjunction with the prime contract. It is only logical that as each prime contract contains terms and provisions relating specifically to a construction project, the subcontracts should take into account the specifics of the prime contract and the project as well. Unfortunately, many contractors rely on and use the same subcontracting contract, over and over, to do all the subcontracting on their jobs. Although this practice may give a contractor a sense of confidence about the contents of the subcontract it signs and save some attorney's fees in the process, it can have disastrous results. The more informal nature of a private project may tolerate the use of such standardized subcontracts because of the ability of the parties involved to be flexible and adjust to any problems that arise. The formal nature of public works construction contracts, however, does not allow such flexibility or adjustment.

Time for Performance

If the timing of a project or a particular part of a project is critical under the terms of the prime contract, it will be necessary to coordinate the terms of the subcontracts to reflect that fact. When a time requirement for a subcontractor's performance is set forth in a subcontract, it becomes a contractual obligation for which the contractor can seek damages if breached. If the subcontractor failed to meet the required schedule, the contractor could then seek recovery for damages suffered as a result of the subcontractor's breach. This could become very important if the prime contract contains a liquidated damages clause. From the viewpoint of the subcontractor, it is equally desirable to obtain schedule commitments from the general contractor as well.

Although there is no magic remedy for insuring that the equipment or material suppliers will meet their contractual obligations to deliver by a committed date, it is important to include statements in agreements with such suppliers that set forth the time constraints of the project and the damages that will be suffered in case of late delivery. This will set the stage for recovery through legal action if the supplier defaults on its delivery schedule.

Incorporation of Prime Contract

The prime contract outlines the responsibilities, duties, and liabilities of the general contractor. As the general contractor will contract with subcontractors to accomplish the work called for in the prime contract, all applicable terms, conditions, and provisions of the prime contract relating to that portion of the work to be done by the subcontractor must therefore be included in the terms of the subcontract itself.

Ideally, a contractor should study the prime contract and carefully draft a custom subcontract for every subcontractor who will be working on a project. If the time and expense to prepare such custom subcontracts are not warranted, a single all-purpose subcontract to be used for all subcontracting on the project may be very useful. It will ensure that all subcontracts work together and contain the provisions required by the prime contract. With modifications made as needed for the various subcontractors, an all-purpose subcontract is almost as good as a custom contract.

It is often a requirement of the prime contract that its terms be incorporated into each such subcontract, and they are usually incorporated by reference. Whether required or not, it is generally a good practice to do so as it guarantees that the subcontracts will be coordinated with the prime contract and each other. However, the incorporation of the prime contract does not in itself make for good subcontracts. The individual subcontracts must still be reviewed to ensure that the incorporated provisions of the prime contract make sense under the circumstances involved, and that the subcontracts reflect the specific work and liabilities to be undertaken on the projects.

If the prime contract requires the incorporation of its terms in all subcontracts, the contractor may want to make a similar requirement a part of all its subcontracts, should any of the first tier subcontractors do any further subcontracting. By requiring the incorporation of these terms, the general contractor can be assured that all of the subcontractors on the job are operating under the same set of rules. More efficient project administration is usually the result.

As a practical matter, if the terms of the prime contract are incorporated into the subcontracts, each subcontractor may be entitled to receive a complete copy of the prime contract. In many cases, the documents that make up the prime contract are quite voluminous, and producing copies of them for all of the subcontractors may prove burdensome to the general contractor. If the prime contract requires it to be incorporated in full in every subcontract, the general contractor must be prepared to supply all of the documents needed or requested by the subcontractor. If the contractor itself elects to incorporate the prime contract into its subcontracts, the burden of providing copies of all of the contract documents can be avoided by incorporating only those parts of the prime contract that need to be included.

Key Provisions for Subcontracts

Whether the terms of the prime contract are incorporated or not, contractual provisions to be included in the subcontract must be viewed in light of how they are addressed in the prime contract. The contractor must do everything possible when drafting its subcontracts to ensure that they will coordinate with the prime contract.

The following list of topics or contract provisions is by no means intended to be exhaustive or to serve as a checklist for the preparation of any subcontract. Rather, it is a list of subjects that need to be coordinated between the prime contract and subcontracts, as they are often causes of controversy that can be avoided if properly addressed in the subcontracts.

Indemnity Provisions

As discussed in Chapter 3, *Risk Allocation,* contractual indemnity provisions, as permitted in many states, include provisions that not only require the contractor to indemnify the agency and its agents for claims and losses, but also to defend and hold them harmless as well. With the ever-increasing cost of defense and the costs of construction claims, indemnity provisions in most prime contracts have become more and more onerous for the general contractor. It is not at all uncommon for a prime contract to require the contractor to indemnify, defend, and hold harmless the project engineer or architect, as well as the public agency, from any and all construction claims unless the claim or loss is the result of the *sole* negligence of the architect-engineer or the agency.

If the agency's sole negligence is the cause of a loss or claim, the contractor and its insurance company will still have to provide the agency with a defense until the fact of sole negligence is proved. In the meantime, the contractor and its insurance company are responsible for the cost of that defense.

One way for the contractor to lessen its indemnity burden is to transfer the indemnity obligation to its subcontractors, as discussed under "Reallocation of Risks to Subcontractors" in Chapter 3, Risk Allocation.

Insurance and Bonds

Just as prime contracts generally require the contractor to obtain specific types of insurance and bonds, in designated amounts of coverage, the subcontracts should contain requirements that the subcontractors obtain and show proof of securing the same insurance to cover their work on the project. The contractor should consult with its insurance broker to ensure that the type and amount of insurance required and supplied by the subcontractor is satisfactory. Setting insurance requirements is one of the easiest and safest forms of risk allocation a contractor can use.

Whether it is a provision of the prime contract or not (and it normally is), any insurance policy taken out by a subcontractor should list the agency and its agents as an also-named insured. In addition, the contractor should always be listed as a named insureds. By being named on the policy, the contractor as well as the agency and any other listed additional insureds have the right to make a claim directly against the policy should there be a loss, without having to go through the subcontractor.

Prime contracts generally specify when both the contractor and the subcontractors will be required to supply performance and payment bonds, and when only the general contractor will be required to do so. In the past, bonds were only required of the general contractor, but recently, subcontractors are being asked to provide bonds more frequently. If it is a requirement of the prime contract that the subcontractors be bonded, it is imperative that the subcontracts, whether by the incorporation of the prime contract by reference or otherwise, include bonding requirements. The contractor must also ensure that any such bonds provided will meet the requirements of the contract.

When subcontractor bonds are not required by the prime contract, the contractor may still find them useful. If the contractor is uncertain of a particular subcontractor's ability to perform, it may require the subcontractor to post a performance bond in favor of the general contractor. The general contractor will most likely have to pay the premium on such a bond out of its own pocket, but it can be well worth it if the bond requirement results in a higher quality subcontractor or completion of a job after a subcontractor's default.

Audit Requirements—Record Keeping

Projects that are funded by an agency normally include audit provisions that require the contractor to maintain certain records for an eventual audit. If the contractor is obligated to maintain audit records, the subcontractors must be obligated to maintain them as well. Any failure of a subcontractor to maintain such records will be the responsibility of the contractor.

Compliance with Special Laws and Regulations

When a prime contract requires the contractor to comply with special laws or regulations, such as an affirmative action or "Buy American" program, the subcontracts must reflect this requirement. A subcontractor's failure to comply with such requirements could cause the general contractor a great deal of trouble and possibly cost a great deal of money.

Guarantees and Warranties

Very often on larger projects, the public agency will take over the use of a part of the project prior to final completion. Technically, warranties or guarantees begin to run as soon as any installed equipment is put into service. However, it is not uncommon for the prime contract to contain language extending the warranty and guarantee periods. Any such extension should be disclosed to the subcontractor in the terms of its subcontract.

PRECONSTRUCTION CONFERENCE

The appropriate time for the *preconstruction conference,* or construction coordination conference as it is sometimes known, is after all of the subcontracts have been awarded but before actual construction begins.

It is important that all key members of the construction team be represented at the meeting. The presence of a representative of the public agency will enable the agency to better appreciate the operational problems that may be encountered by the construction contractor, will aid the entire construction team by providing a better understanding of the agency's needs, and will help the architect-engineer obtain a quality project that is consistent with the planned schedule and estimated construction cost.

The preconstruction conference is the time that potential construction problems can be discussed and possibly prevented. It is designed to benefit all concerned by recognizing the responsibilities for various tasks *before* a project is begun. The benefits include the following:

1. Recognition and elimination of delays and disagreements.
2. Establishment of agreements that control increases in construction costs.
3. Early resolution of gray areas of responsibility that, if left unresolved, can cause later disputes.
4. Early contact with utility owners with whom coordination is necessary.
5. Establishment of the authority and responsibility of the parties to the contract.
6. Discussion of the procedures for handling of changes, payments, submittals, and job safety and security.
7. Discussion of permit requirements and the influence of other public agencies and utilities.
8. Establishment of requirements for temporary facilities and controls and responsibilities for the property of others.
9. Identification of easements and rights-of-way, staging areas, disposal areas, and traffic limitations.

Definitions

The preconstruction conference, or construction coordination conference, is a meeting of the principal parties involved with the planning and execution of a construction project and should include

A representative of the public agency/owner.

The architect-engineer and its resident project representative.

The general contractor and all prime contractors and their subcontractors.
The subcontractors and their superintendents.
Key suppliers.
Public regulatory agency representatives, as necessary.

Full attendance and participation by all key members is often required under the terms of the General Conditions of the Contract or other contract documents to assure that the general contractor and all subcontractors attend the conference.

Purpose

The primary purpose of the conference is to establish acceptable ground rules for *all* parties concerned and to assure that each contractor understands the complete job requirements and coordinates its work to produce a completed job in a minimum amount of time with maximum profitability and in a spirit of cooperation with the public agency, the architect-engineer, prime contractors, and all subcontractors and suppliers.

Time for the Conference

The preconstruction conference should be scheduled to permit sufficient time to cover the total agenda. This could involve a meeting of from one to several days. In any case, whatever time is spent should be considered *preventive* rather than corrective.

Frequently, there may be two such conferences. The first conference is between the agency representatives and the architect-engineer to review the proposed agenda and establish a position on each of the issues listed in the agenda. The second conference is the one that the contractor is expected to attend.

Subjects for Discussion

The subjects to be discussed during a preconstruction conference depend upon the nature, size, and complexity of the project. It is, however, necessary to assign priorities to the tasks involved. Although each job is different, certain factors are common to all types of construction. As the agenda is normally prepared by the architect-engineers, a preconference contact with them will give a contractor an opportunity to suggest subjects for the agenda, to assure that these subjects are addressed.

Agenda Items for a Typical Preconstruction Conference

1. *Progress Payments* When, how, and to whom in *exact* terms, stated clearly so that no question exists about requirements and responsibilities. This should also apply to the area of retainage and final payment.

2. *Definition of Substantial Performance* Its impact on contract time, liquidated damages, insurance, punch list items remaining, final payment time, and release of retainage.

3. *The Form of Payment Requests* Identify the form that must be used for the submittal of progress payment requests. Determine if lien waivers are required from suppliers and subcontractors. Can supplier and subcontractor waivers be one month behind? That is, can they be submitted the following month, after the general contractor has had a chance to pay its suppliers and subcontractors?

4. *Payroll Reports* Requirements, if any, for certified payroll reports (typical on federally funded projects).

5. *Shop Drawing and Sample Submittal Data* The form and procedure for the shop drawings and samples. Procedure for handling subcontractor submittals. Identify the parties authorized to receive submittals; where they must be delivered; number of copies of each required; turnaround time to return submittals to the contractor; precedence of contract drawings over shop drawings; types of architect-engineer approval or disapproval, that is, "approved," "approved as noted," "revise and resubmit," "rejected," and so on.

6. *Insurance; Permits Required* Time for the submittal of insurance and obtaining permits. Who obtains and who pays for permits.

7. *Job Progress Scheduling* A preconstruction conference provides an opportunity for the essential involvement of subcontractors in the development of and coordination of the individual schedules that make up the overall construction schedule for the project. Many large projects are conducted on an overall schedule using the CPM or PERT systems. Such scheduling, however beneficial, is useless unless understood and used by all parties.

8. *Temporary Facilities and Controls* These are the utility and other services that are essential to the construction process, but do not form a part of the finished project. Under the CSI Format, these requirements are spelled out in Division 1, General Requirements,

of the technical section of the specifications. This is often a gray area requiring clear definitions of responsibility.

Such services include temporary environmental controls such as dust abatement, air pollution control, noise abatement, rubbish control, sanitation, toxic wastes, and the use of explosives and blasting; protection and restoration of existing facilities or improvements; and temporary power and light, temporary water supply, temporary street use, contractor's work and storage area, transportation facilities, street closures, safety measures, and access to the work site.

Some questions relating to the services just mentioned that must be answered are

(a) Who provides the services?
(b) Who maintains them?
(c) Who pays for them and in what proportion?
(d) What are the *contractual* responsibilities, *if any*, under the provisions of OSHA?

10 *Jobsite Security During Nonworking Hours* Losses from vandalism and theft at unguarded construction sites are a rapidly mounting source of expense to the entire construction industry. An agreement sharing the costs of better security measures would be well worth discussion.

11 *Cleanup and Trash Removal* Boxes, scraps, food containers and wrappers, sanitary wastes, and the like.

12 *Available Hoisting Facilities* Who supplies the hoisting facilities? If the general contractor supplies them, what will the arrangement be in order to make the hoist available for the individual subcontractors?

13 *Change Orders* Because change orders are the subject of more disputes than any other single aspect of a construction project, they should be discussed in complete detail. Typical items for discussion include

(a) Percentages of overhead and profit.
(b) What costs will or will not be allowed.
(c) Length of time a Change Order price is firm.
(d) Identify individuals authorized to issue.
(e) Procedures for Change Order Proposals.
(f) Emergency change procedure.

(g) Change Order forms that must be used.
(h) Cost isolation documentation for changes.
(i) Time extensions—requests by subcontractors.
(j) Impact costs.
(k) Complete breakdown required on subrequests.
(l) Overtime due to Change Orders; productivity.
(m) Who owns material or equipment removed.
(n) Who disposes of removed items.
(o) Responsibility for as-built drawings due to changes.

14 *Punch List Procedures* See Chapter 12, Project Closeout, for recommended punch list procedures.

It is of considerable importance to establish meaningful communication between the parties involved on a construction project.

NOTICE TO PROCEED

The accepted procedure in public works contracting is a four step process beginning at the opening date, as illustrated in Figure 8.1. Typically, the agency after opening bids will examine them and tabulate all of the bids on a spread sheet. An analysis will be made of unit price bids to determine if any serious imbalance exists. After the evaluation and determination of the apparent low bidder, a recommendation will be made by the agency's engineer and the matter will be referred to the agency's governing body for award.

Upon deliberation by the governing body of the public agency, an award will be made to the lowest responsive, responsible bidder. This may not be in the form of a letter to the successful contractor but may simply be an entry in the minutes of the meeting of a city council, board of supervisors, board of directors, or a similar governing body. At the time the award is made, if there are, say, ten bidders, the agency will probably hold the bid bonds of the three lowest bidders pending the low bidder's signing of the agreement. This is just in case the low bidder defaults or becomes ineligible for one reason or another.

This is the time, in most public contracts, when the low bidder must obtain all bonds and insurance and present them to the agency along with a signed agreement document for the agency's signature. Remember, no contract exists as a result of an award, and access to the site may still be denied. After approval of the bond forms and insurance, the agency will then sign the agreement, and a contract exists.

Although some contracts state that the work must begin within ten days after the date of the execution of the agreement, a contractor has a good argument for extra time where the receipt of the signed copy of the agreement is sufficiently delayed so that it impacts the time that would have otherwise been available to the contractor. After all, there is no reason why a contractor should move onto the site until it actually sees the signed agreement. If the agreement is being mailed to you, the agency will probably assume that the dates in the agreement are reasonable. The preferable procedure is to follow the agreement with Step 4 (in Figure 8.4) by the issuance of a Notice to Proceed. In this manner, the starting date is fixed, and there can be no argument later in the project as to the exact date of completion of the work.

On federally funded projects there is usually a significant delay period between the time of executing an agreement and the issuance of a Notice to Proceed in order to allow the final funding arrangements to be made. On EPA work, this delay is often 60 days' time during which time the contractor's bid must be firm. On a HUD project some years ago, the author was involved in a 90 day delay for funding.

SCHEDULE OF VALUES ON LUMP SUM JOBS

One of the tasks disliked most by contractors new to the public sector is the demand for submittal of a bid breakdown, or *schedule of values* as it is called, on a lump sum job. If required under the terms of the contract signed by the contractor, it is an obligation, however, and whatever the cost, the contractor is required to make the submittal. The format is exactly like that for a unit price project, complete with individual line items, an estimate of the quantity of each line item, the unit price assumed for each such item, and the extension or product of the unit prices times the estimated quantities.

Some contractors have tried to say "it's none of the agency's business." But they are wrong. If it is in the contract requirements, then that is the way the job was bought, so the contractor has no other choice but to live up to the contract's obligations. Remember, the only legitimate purpose in the preparation of a *schedule of values* is to allow the agency to verify the monthly payment amounts of progress payments. What a contractor does have a right

SCHEDULE OF VALUES ON LUMP SUM JOBS

to object to, however, is any attempt by the agency to utilize the unit prices named in the schedule of values *for any purpose other than estimating monthly progress payments.* Schedule of value submittals are *not* valid for pricing change orders. Don't worry about it . . . . On a lump sum contract, a contractor is entitled to the full bid amount as amended by any change orders, regardless of the distribution of costs on the schedule of values. Avoid extra trouble, however, and be wary of preparing a badly unbalanced distribution, as it may have to be redone a few times before its acceptance by the public agency.

CHAPTER · 9 ·

Project Administration

- The Contract
- Administration
- Documentation
- Temporary Facilities
- Scheduling and Coordination
- Contract Time
- Differing Site Conditions
- Change-Order Administration
- Job Changes and Performance Bonds
- Labor and Material Releases
- Payment
- Materials and Methods

CHAPTER 9   PROJECT ADMINISTRATION

THE CONTRACT

Contract Documents

It should be borne in mind that on public works projects, there are numerous documents considered a part of the *contract documents* that on most private projects are provided for general information only, prior to bidding. This essentially means to the contractor that any of the terms and conditions outlined in each of the contract documents becomes a binding requirement on the contractor, and every word, phrase, or statement contained in these documents is a binding requirement of the construction contract. Typically, in public works contracts, the contract documents may include the Notice Inviting Bids, the Instructions to Bidders, the Bid (or Proposal), the Bid Sheets, List of Subcontractors, List of Equipment Suppliers, the Bid Bond, Wage Rates, Agreement, Payment Bond, Performance Bond, Noncollusion Affidavits, Certifications that may be required, General Conditions of the Contract, Supplementary General Conditions, Technical Specifications, Addenda, Project Drawings, Standard Drawings, Change Orders, and in some cases, supplementary material required by a funding agency.

It becomes quite important, then, to carefully review and study *all* of these documents prior to the submittal of a bid, and again prior to the actual start of construction. Each member of the contractor's management team should be intimately familiar with their provisions.

Prescriptive vs. Performance Specifications

Often, references are made to specifications sections as *prescriptive specifications* or *performance specifications*. Although this at first may appear to be more of an exercise in semantics, the difference in these two terms may assume great importance to a contractor, particularly when it may be recognized that a particular section of the specifications contains elements of *both* types.

By definition, a performance specification may be defined as one in which the specifier states the objectives to be met by the contractor, without specifying the detailed method of construction. An example would be a mechanical system for which the specifications spell out the service conditions, equipment capability, method of operation of the equipment or system, capacities, efficiency requirements, or similar attributes of the system. Such a specification could legitimately include detailed requirements about the quality of the basic materials used in the system's construction, such as ASTM standards for various metals, fasteners, or other materials.

A prescriptive specification is one in which the specifier calls out the details of construction to be followed by the contractor. An example would be asphalt concrete paving as it is often specified by the various state highway specifications. In such specifications, the contractor is directed as to what materials to use and how to batch, proportion, mix, transport, spread, and compact the asphalt concrete materials. In such a specification, it is common to find that the requirements for the compaction equipment are specified in detail, including the number of axles, wheel sizes and types, weight of equipment, tire spacing, tire pressures, operating weight per tire, and similar requirements. Following that, the contractor is directed how to use this equipment, stating the number of passes in either direction, the amount of overlap between the passes, rolling pattern, direction of rolling, temperature of mix at the time of rolling, and the final dimensional tolerances of the finished pavement.

The important distinction is that when an engineer or architect writes a specification that tells the contractor not only what the performance requirements are, but the exact details of how it must be constructed, it is equivalent to having the engineer or architect design the work, then asking the contractor to guarantee that the specified methods will work. The courts hold that a specifier cannot specify a *method* of construction in combination with a *performance* requirement stating what the end product must be capable of doing, because it is equivalent to trying to force the contractor to guarantee the engineer's or architect's work.

Either the contractor must build the work according to the methods directed by the architect-engineer, in which case the end performance, or ability to do the job it was designed to do, becomes the responsibility of the architect-engineer; or the contractor, after being advised of performance requirements, must be left to its own methods to achieve that end. In the latter case, the contractor bears the entire responsibility for the satisfactory performance of that product.

Under a strict performance specification, the architect-engineer has no authority to reject a product until it has been demonstrated that it is incapable of performing or has failed to perform in accordance with the terms of the specifications. It may be possible, however, that an engineer or architect, who has reason to believe that a particular construction method is incapable of meeting the specified performance requirements, may be within its contractual rights to require a contractor to demonstrate a proposed method in a test prior to authorizing its use in the project. Under such conditions, the cost of the test, if not a part of the project itself, may be borne by the agency if the contractor's proposed method proves to be acceptable. Often, such tests are conducted by designating a part of the project as a test area and observing the demonstration under close inspection. As such, no time is lost or added expense incurred if the results are satisfactory.

ADMINISTRATION

Supervision of the Work

It is particularly emphasized in public works construction that the contractor must keep a project superintendent or project manager on site at all times while work is being performed. Failure to do so may jeopardize a contractor's monthly progress payments and may negatively impact related delay claims. It is not sufficient to leave only foremen or trade supervisors in charge; a general superintendent or project manager must be designated to the agency at the time of the preconstruction conference, and in his or her absence, an authorized alternate must be provided.

One-to-One Concept

As you may have heard that too many cooks in the kitchen will spoil dinner, so it is with the administration of a construction project. A project should be structured so that the contractor and the agency or their designated engineer or architect are each represented by a single individual. It is essential that this person be designated at the beginning of the job, and that sufficient authority be delegated to that person so that he or she may commit their employer to an obligation.

The contractor should demand that the agency or the architect-engineer designate a single individual through whom *all* orders, directions, or other official communications are transmitted to the contractor. Failure of the agency or its architect-engineer to do this may result in numerous delays in the work, constructive change situations or other circumstances that justify a contractor's filing of claims for the recovery of time or money. Generally, a contractor should be entitled to recover for any delay caused by the agency's administration of the contract.

The appointment of individuals to transmit and receive job information is what the authors call the *one-to-one concept* (see Figure 9.1). Under this

ADMINISTRATION

FIGURE 9.1 A one-to-one relationship is an important element in assuring good contractual relationships.

arrangement, the agency and its architect-engineer agree to transmit all official communications to the contractor's superintendent through its on-site resident project representative, thus eliminating the risk to the contractor of unauthorized actions communicated by parties without prior authority to do so. It also means the designation by the contractor of a single, on-site management person to whom the issuance of an order in the field is equivalent to legal service of notice to the contracting firm.

The Contractor and Subcontractors

Almost all of the construction contract General Conditions are based upon having the resident engineer, inspector, or architect-engineer representing the public agency deal solely with the general contractor, not directly with subcontractors, material suppliers, or fabricators. The General Conditions generally state that the general contractor is fully responsible for all of the acts and omissions of its subcontractors, and nothing in the General Conditions is intended to create a contractual relationship between any subcontractor and the public agency or architect-engineer, or any obligation on the part of the agency to assure that the contractor has paid its subcontractors or material suppliers.

The fact that only the general contractor is recognized should end the frequent disputes of subcontractors that revolve around definitions of the scope of their portion of the work (usually, the alleged result of the general contractor's failure to provide the subcontractors with a complete set of specifications and drawings). The general contractor often complains that this is the architect-engineer's fault because it failed to include certain work items in a specification section that would be performed by a specialty subcontractor, and thus additional funds should be paid to cover the added charges made on the general contract by the specialty subcontractor.

This approach can only get a contractor into trouble: Just one contract was let; the total scope of the work was specified; and it is not the responsibility of the public agency or architect-engineer to determine how the successful bidder plans to subcontract the work. Furthermore, the scope of any one class of work can often vary significantly even from one county to the next due to differences in trade union contracts and the resultant jurisdictional agreements. One contract means one job; unless the contract as written prevents it, it is the general contractor's responsibility to properly contract with its subs to ensure a clear understanding of the scope of each such subcontract.

Submittals of Shop Drawings and Samples

The shop drawing is the connecting link between design and construction. Because of the increasing complexity of today's construction, and often the inexperience of design engineers and architects with construction contract administration, shop drawings in recent years have become one of the largest sources of professional liability claims against architects and engineers. Unreasonable delay in processing shop drawings and ambiguous wording in the shop drawing stamp have been two principal sources of trouble. Most specifications require that the contractor refrain from ordering materials or equipment until approval of the shop drawings. Any delay by the architect-engineer or public agency in processing shop drawings affects the contractor's scheduling and, in turn, may justify the filing of delay claims by the contractor.

Normally, the contractor is obligated to submit a preliminary schedule of the submittals of shop drawings, samples, proposed "or-equal" products, or other submittals required under the terms of the Contract Documents. This schedule is normally reviewed by the architect-engineer and finalized prior to the actual start of construction. Shop drawing submittal procedures is one of the topics that should be discussed at the *preconstruction conference*. Careful attention to these preliminary matters can avoid costly misunderstandings at a later date, and an agreed upon procedure should be set forth in writing, with copies circulated to all parties concerned.

One of the most important and misunderstood facts about shop drawings is that a shop drawing approval does *not* authorize the contractor to change or otherwise depart from the contract drawings or specifications. [Appeal of *Whitney Brothers Plumbing & Heating, Inc.*, ASBCA No. 16876, 72-1 BCA 9448 (1972)] Such a departure or change may only be accomplished by a change order. Many field superintendents firmly believe that when a detail has been shown differently on a shop drawing than on the original contract drawings, the shop drawing will take precedence. This is a fallacy created by the belief that the agency's or architect-engineer's approval of a shop drawing means carte blanche acceptance of everything that is contained on the drawings. Shop drawing approval is intended only to determine that the shop drawings are in conformance with the basic design concepts of the project and in compliance with the information provided in the Contract Documents. The contractor is still responsible for the detailed dimensions to be confirmed and correlated at the jobsite; for information that pertains solely to the fabrication processes or techniques of construction; and for coordination of the work of all trades.

Therefore, a shop drawing should not be considered a change order, and any variation from the design drawings and specifications must be the result of a change order. Otherwise, it is not authorized, and you may not get paid. On the other hand, if the architect-engineer improperly returns a shop drawing with design changes added, the contractor is in a strong bargaining position for an extra. As the architect-engineer may have just created a *constructive change,* the contractor would then not only be entitled to an extra for such a change, but also a claim for any delay that is caused.

The shop drawing submittal process is beset with risks for all parties. However, if a contractor wants to protect itself, it should follow a few simple rules.

1. Submit and resubmit shop drawings on time.
2. If revisions are required, do them promptly.
3. Do not start work covered by shop drawings until approval of the shop drawing submittal or you may be required to demolish the work and rebuild.
4. If you are proposing a change, clearly mark the change on the shop drawing. *The general rule is that if a contractor explicitly and promi-*

nently calls out any deviation from the plans and specifications that are contained in the shop drawings, the agency will be bound by the approval regardless of the wording of the exculpatory clause on the shop drawing approval stamp.

5   If the architect-engineer or agency delays excessively in returning shop drawings after review, you may be entitled to a delay claim if your activities are clearly documented.

6   Keep a submittal log of all shop drawings, samples, or other submittals. Document the dates transmitted to the architect-engineer and the date of return. List, also, the action taken on each, along with other pertinent data (see Figure 9.2).

One thing the contractor should remember is this: Keep the deck stacked in your favor. There are many elements involved in doing this. The submittal log is one important element; however, a submittal control sheet (see Figure 9.3) better allows a contractor to keep on top of the problem of delays. If there are going to be delays in the submittal process, don't let them be yours. Another important step is to be responsive to all of the contract requirements. Many a contractor has jeopardized its advantageous position on a claim because of complicity in delays caused by the contractor's failures to submit as specified.

Inform all subcontractors that their submittals must be received in advance of the date you are required to submit to the public agency or architect-engineer. Before submitting to the agency or architect-engineer, however, these documents should be reviewed carefully for compliance and then each copy should be stamped and signed before submittal. The contractor should be aware of the importance of handling shop drawings on its own end. The contractor should not submit any shop drawings, samples, or any other submittals to the agency or architect-engineer without reviewing them first. Then, the contractor should have a rubber stamp made, similar to those used by the architect-engineer, which can be used to identify those shop drawings that have been reviewed.

Equally important, the contractor should be sure that the wording of the stamp contains a statement that the contractor has reviewed these documents and has approved them prior to submittal to the agency or architect-engineer. This is quite important because there is no direct contractual relationship between the agency or architect-engineer and any of the subcontractors. As a result, the agency or architect-engineer would be vulnerable to claims for extra work or charges that it authorized work that was

E. R. FISK CONSTRUCTION

CONTRACTOR SUBMITTAL LOG

Project: 5 MGD Treatment Plant
Job No.: 87-5463
Project Mgr.: R. E. Barnes
A&E: LFM Engineers, Inc.

Date Rec'd	Transmittal No.	Description	Subcontractor / Ref. Spec. Section	A&E Trans. No.	No. Copies	Action: No Exceptions Taken	Action: Make Corrections Noted	Action: Revise & Resubmit	Action: Rejected	Date Ret'd	No. Copies Ret'd	Remarks
6-11-87	75	Plan View – Aeration Basin Secondary Sedim. Tanks Dissolved Air Flot. Thk.		75	6					6-22-87	2	
6-12-87	76	Electrical Materials List	A.J. Peterson / 16A	76	6	✓				7-17-87	2	
6-12-87	76	Warranty covering Labor & Matls for Acme Kitchen Unit	/ 10D	77	6		✓			6-16-87	2	
6-12-87	76	Metal Compartment Dwgs. Color Card	/ 10A	77	6		✓			7-20-87	2	
6-12-87	77	Celltite Resin System	/ 3C	78	6			✓		6-19-87	2	
6-12-87	77	Pumps, Flow Detector Motors, Frequency Drive	Harris & Foote / 11B	79	6	✓				7-20-87	2	See Change Order C.O. 001
6-12-87	77	Chlorination Equipt.	Wallace & Tiernan / 11I	80	6				✓	6-18-87	2	
6-12-87	77	Curing Compound	/ 3C	81	6		✓			6-19-87	2	Approved as an "or equal"
6-19-87	78	Wall Spool Blower Solids Bldg.	Data used are fictitious	82	6		✓			6-22-87	2	
6-22-87	79	5. Primary Sludge Digester Slab	for illustration only	83	6		✓			6-23-87	2	

Contractor Submittal Log provides a permanent record of all submittals by the contractor of shop drawings, samples, and other requested data received during construction.

Wiley-Fisk Form 7-2

FIGURE 9.2 Shop drawing and sample submittal log.

SUBMITTAL CONTROL SHEET

E. R. FISK CONSTRUCTION

Project Title: Pump Station, Sewer, & Drainage
Project Manager: R.E. Barnes
Project No.: 88-4597

SECTION NO.	ARTICLE NO.	DIV.	SPECIFICATIONS SECTION TITLE (Indicate Division No. if applicable)	SAMPLES	SHOP DWGS.	MAT'L OR PARTS LIST	DESCRIPTIVE DATA	MFRGS LITERATURE	MIX DESIGNS	CERTIFICATES	OPERATION INSTR.	TESTS	DATE OF SUBMITTAL	DATE REJECTED	DATE RESUBMITTED	DATE ACCEPTED	NOTES
2E	1.04	2	Chain Link Fencing	✓	✓								2/17			2/27	
2H	1.04	2	Piling		✓					✓		✓	2/25			2/28	
3A	1.04	3	Conc. forms & Falsework		✓								3/5			3/19	
3B	1.04	3	Conc. Reinforcement		✓							✓	3/5			3/16	
3C	1.04	3	Cast-in-Place Concrete		✓				✓	✓		✓	3/5	3/15	3/28	4/9	Subs. Additive
5A	1.04	5	Struct. Metalwork		✓								4/27			5/16	
5B	1.04	5	Misc. Metalwork		✓								4/27			5/12	
7A	1.04	7	Calking & Sealing	✓						✓			3/15			3/27	
9A	1.04	9	Painting & Protec. Coating	✓	✓							✓	6/20			7/6	
11A	1.03	11	Diesel Engine & Appurtenances		✓	✓	✓	✓		✓	✓		7/8	7/22			
11B	1.03	11	Diesel Engine-Generator		✓	✓	✓	✓		✓	✓		7/8	7/22			
13A	1.03	13	Pre-fab Metal Bldgs		✓								5/18	9/2			
14A	1.04	14	Trash Rack Hoist System		✓							✓	6/12			6/30	
15C	1.03	15	Gates, Valves, & Appurtenances		✓	✓											
15D	1.03	15	Pumping Equipt.		✓	✓	✓	✓		✓	✓						
15E	1.02	15	Portable Fire Extinguishers					✓									
15F	1.04	15	Hydro-Pneumatic Tank		✓		✓	✓		✓							
15G	1.03	15	Well Pump & Motor		✓	✓				✓							
16A	1.06	16	Electrical Work		✓	✓				✓	✓						
17A	1.04	17	Controls & Instrum.		✓	✓				✓							

Data used are fictitious for illustration only

Wiley-Fisk Form 7-4

FIGURE 9.3 Submittal control sheet.

beyond the scope of the construction contract by approving shop drawings that did not contain the contractor's acceptance stamp. Thus, what the agency or architect-engineer needs are submittals from the general contractor, not the subcontractors.

By stamping and signing the subcontractor submittals, they become submittals from the general contractor. The public agency or architect-engineer is within its rights to return unsigned submittals without action and to require that they be reviewed and signed before resubmittal. The problem with not following the proper submittal procedure is that it causes delays to the contractor on the job. Worse yet, such delays are unrecoverable, as they will be interpreted as having been caused by the contractor. If the contractor misses its project delivery date as a result, there may even be an added cost burden in the form of liquidated damages assessed by the public agency for the contractor's failure to complete the work on time.

Again, *technically,* only a change order can authorize a deviation from the Contract Documents. But remember, a constructive change has the same effect, and the price is open to negotiation.

DOCUMENTATION

From the standpoint of good business practices, there is probably no subject more important than competent documentation. Documentation is not just a means of keeping business records for cost control; it is also a way of providing a meaningful audit trail in support or defense of construction claims. The party with the best documentation stands the best chance of winning in a claims situation.

Records and documentation play probably the most important role in the successful settlement of construction contract claims. The daily events and details of a job must be documented to substantiate claims and prove damages. Facts must be recorded and preserved. Armed with a carefully prepared claims package of facts and figures, a contractor can support its position and move negotiations toward a favorable settlement.

All too often, however, contractors do not keep good records. Procedures may be set up properly and the files filled with documents, but frequently the details of daily events are inadequate or not readily accessible, if available at all. This is understandable, as most construction personnel are more concerned with construction than with keeping records, but it is also potentially damaging to the contractor.

CHAPTER 9 PROJECT ADMINISTRATION

Records and Record Keeping

Most contractors know that it is important to keep good business records but do not find out just how important it really is until records that should have been kept, but were not, are sorely needed. A contractor must keep the same types of financial records as any for-profit enterprise to meet a variety of business and management purposes. In addition to records required by law and the records that various governmental agencies require regarding such things as taxes, payrolls, and similar business records, contractors must keep records that will serve as source material for obtaining indispensable support services. Financial statements and reports are required to obtain a loan or a credit line at a bank, a bond from a surety, an insurance policy from an insurance company, and a project from a public agency. These particular records are especially important to a contractor, and without them, a contractor will be unable to grow and take on even larger, and hopefully, more profitable projects. A contractor should develop a *team,* including its bonding agent, insurance broker, banker, construction law attorney, and the like, and should consult with them to assure that the required records are kept in an acceptable manner.

Records Required by the Contract

In addition to normal business records, a contractor is often required to keep particular types of records as specified in a construction contract. For example, on some projects in which federal money is involved, the contract may require a contractor to keep certain records relating to the project. It may even require the contractor to keep such records for a specified number of years and to make them available for a government audit if requested. When an audit function is a part of a contract, the type of records required very often go well beyond summaries and will include all of the contractor's detailed backup information as well. On a large project, the tracking and filing of all such information can become a major administrative task that needs to be carefully considered by the contractor at the time its bid is being prepared.

A contractor is often required by contract to keep certified payroll records and records relating to the contractor's and its subcontractors' compliance with Equal Employment Opportunity (EEO)[1] and Minority Business

---

[1] Equal Employment Opportunity Provisions contained in Title VI of Civil Rights Act of 1964, 42 USCS. §§2000(d), et seq.

Enterprise (MBE)[2] provisions. When such records are required under the terms of a contract, failure by the contractor to maintain such records can be considered a breach of contract. The contractor could then be held liable for damages suffered by the agency as a result of the contractor's breach. The demand by a funding agency for a return of grant funds due to the contractor's inadequate records after an audit is an example of such damages.

Records for Claims and Disputes

In the construction field, disputes are strongly related to the facts. In any arbitration or litigation that may arise out of a claim, the party with the better records evidencing the facts usually wins. With the claims and disputes, not only are the contractor's financial records important, but so is every other type of record. Correspondence, memos, requests for information, change orders, drawings, submittals, transmittal letters, file memos, telephone messages, and much more all become important. A contractor must carefully administer its projects and develop a record-keeping system that will insure all necessary project records are kept and organized properly.

It is normally easiest for a contractor to organize its records on a project-by-project basis and to use standardized forms and procedures whenever possible. By developing forms for requests for information, phone memorandum, daily superintendent forms, accident report forms, and other similar forms containing specific information that has to be filled in by the user of the form, a contractor can be assured that a certain minimal amount of project information will always be documented.

A major portion of any claim relates to damages. For a contractor to recover its damages relating to a claim that has gone either to arbitration or litigation, the contractor will have to prove its damages. Obviously, the better the contractor's records of its damages are, the easier it will be for the contractor to win its case. Whenever a claim or dispute arises, a contractor should immediately take whatever steps are necessary to begin segregating costs and expenses incurred relating to the claim or dispute (cost isolation). For example, if an agency's delay forces a contractor to accelerate the job in order to remain on schedule, the contractor should identify and keep track of all extra costs incurred in staffing the job with extra personnel in order to get back on schedule. Good cost records, developed and maintained by the

---

[2] Minority Business Enterprises Provisions found in Section 103(f)(2) of the Public Works Employment Act of 1977 (91 Stat.1160; see also 42 USCS §6705(f)(2)).

contractor relating to a particular claim or dispute are very persuasive in an arbitration or trial. They are hard to discredit or disprove and usually lead to a favorable result for the contractor.

Types of Records for Claims Support

The types of documentation that should be maintained by the contractor include timecards, payrolls, material receipts, purchase orders, invoices, cash flows, quantity takeoffs, correspondence, forecasts, schedules, subcontracts, and computer cost reports.

All communications should be in writing. If originally expressed orally, any substantive communication should then be put into writing, even if its ultimate destination is only your own office file.

Employee Time Records

In addition to establishing the number of hours that each person works, time records show what each person worked on, whether on straight time or premium time, and the trade involved. In addition, timecards can be used by foremen to record daily production and daily occurrences such as unusual site conditions, weather conditions, delays, accidents, subcontractor problems, or other resource-related details.

Job Costs

The job cost accounting system should follow the original estimate, if the estimate was broken down in sufficient detail. It is desirable that the job cost accounting system be similar in format to that used in the bid estimate. This enables a contractor to compare job progress with the budget as well as serving as a basis of comparison for claims. Although many agencies are reluctant to use the values established in the contractor's bid estimate as a basis for establishing extra costs, sometimes it is the only game in town.

When it becomes necessary for a contractor to justify the cost of extras or claims, it is not enough for it to produce computer reports of unit costs or production rates. What is needed is a valid documentation of production rates and job progress and costs.

Progress Reporting

Often, a public works contract will require the contractor to prepare and submit a daily construction report (see Figure 9.4). Generally, the content of

FIGURE 9.4 Typical contractor's Daily Report form. NOTE: Although commonplace in the industry, this form requires improvement to provide maximum protection. This can be done by adding spaces for documenting all visitors to the site, whom they represent, the time in and out, and remarks.

such a report should include the following:

1. Project data, including project name, location, job number, agency name, engineer name, project manager's or superintendent's name and signature, and the like.
2. Date and consecutive serial numbering.
3. Weather: temperature, wind, humidity, rain, snow, and so on.
4. Number of employees and foremen on site for contractor and all subcontractors.
5. List of visitors to the site, including company affiliation.
6. Heavy equipment at the site and whether operational on that date.
7. Description of work progress accomplished on that date.

A report should be submitted for each and every day of the project, even if no work was performed by the contractor or any of the subcontractors on that date. It is important to account for every day. A report covering a day of no work need only state "no work today: rain" or a similar statement appropriately describing the weather conditions at the site that day.

In addition to the preparation of daily reports, many public agencies will ask for progress charts indicating a graphic comparison of the planned schedule with the actual work progress. The most common of these types of charts is the bar chart (see Figure 9.5) in which the line items correspond to the line items on the bid sheet or schedule of values, although they can represent any division of the work selected by the architect-engineer or the agency.

Another frequently used progress-reporting chart is the "S curve," which is sometimes plotted on a separate sheet, as in Figure 9.6, or may actually be superimposed over the bar chart. The particular value of the S curve chart is that it relates to the progress of the overall project, not just a series of separate line items, and its even greater value lies in the fact that it can be used to predict a scheduling trend. It will show, for example, whether a project is ahead or behind schedule and whether at the present rate of progress it will continue to keep up or move further ahead or behind.

On-Site Relations or Transactions

A part of every contractor's documentation system, irrespective of the demands made by the agency, should be up-to-date construction diaries (see

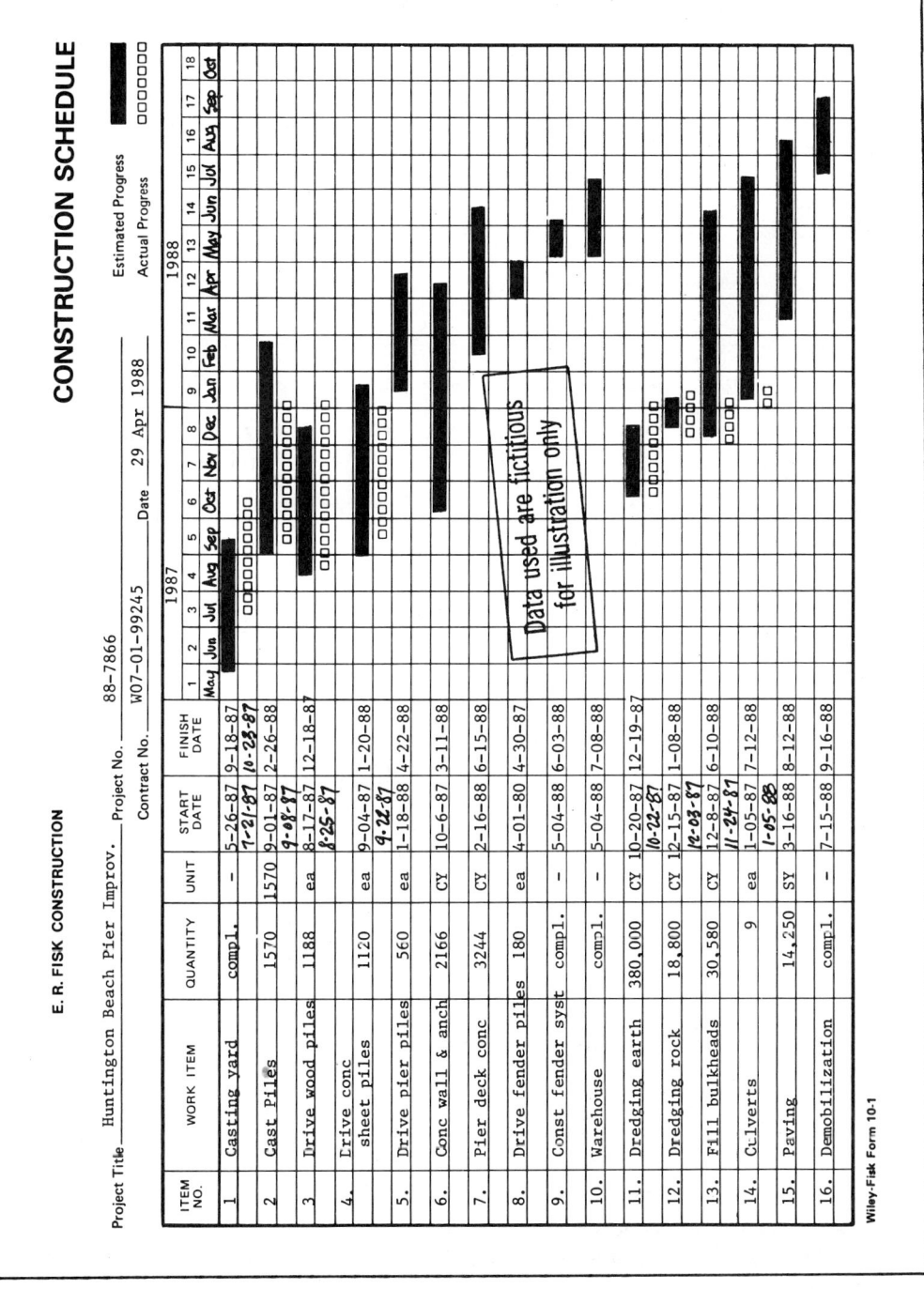

FIGURE 9.5 Construction progress record—bar chart.

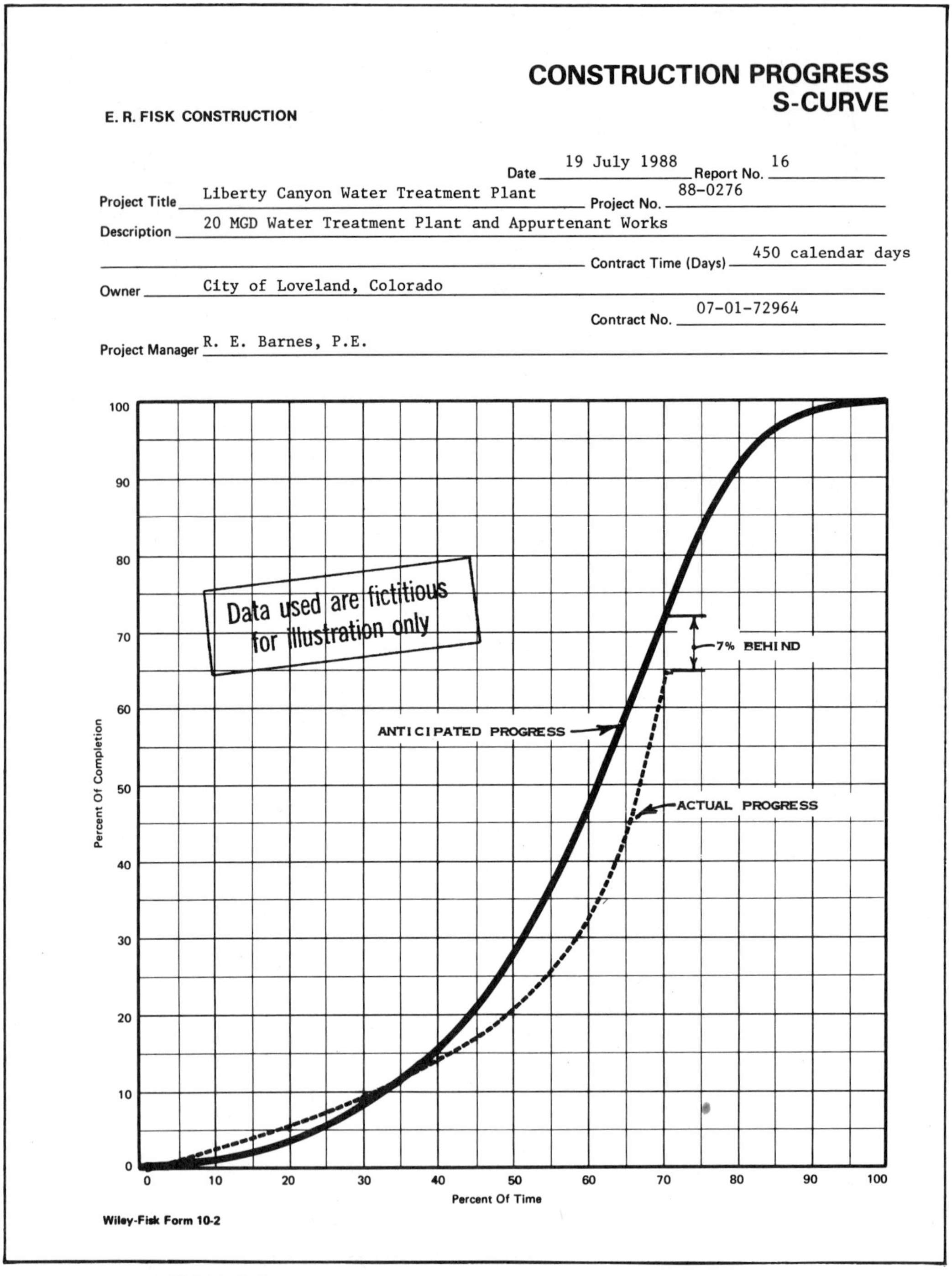

FIGURE 9.6 Construction progress report—S curve.

Figure 9.7), kept by each and every management or supervisory member of the contractor's field crew. The diary should not be a duplicate of the information contained in the daily report, but rather a supplement to that report. The diary should contain notes covering substantive conversations with any agency personnel, the architect-engineer, inspectors, and occasional visitors. It should refer to dates that requests were made of the agency or architect-engineer, any demands made by the inspectors above and beyond requirements of the contract, the substance of all business calls and responses, agreements and commitments made between the parties in the field and a summary of the same, and similar data. Copies of the diary pages, kept in whatever form, should be given to the project superintendent on a weekly basis.

The diary will generally be needed in arbitration or litigation if the keeper is called upon to testify. It can be used by the writer to refresh his or her memory during examination or cross-examination in court or arbitration and possibly entered as evidence. A few basic rules should be followed, however, to ensure that the diary can be used as evidence if needed.

1. Use a stitched bound book.
2. Number pages in ink. Preferably preprinted.
3. Use ink or hard pencil.
4. Do not erase, cross out errors.
5. Do not tear out pages; mark void instead.
6. Report facts, not opinions.
7. Account for every day, even if no work is performed.
8. Report the facts on the *same day* that the incidents occurred.
9. Log record photos taken in diary.
10. Sign and date every entry.
11. *Most important*: *Be consistent* in the way the diary is kept and what is recorded.

Time and Cost Monitoring

A convenient format for the documentation of project time and cost each month is shown in Figure 9.8. The regular maintenance of such records will enable the contractor to be able to respond to agency inquiries promptly, assist in supporting its claims position in many cases, and serve as a check on

> 67
>
> Thursday 6 February 1986   PARKER RESERVOIR
>
> Worked D-9 & 46-A on Ex for backfill on 1½:1 slope on Res. Don't know if we will have enough room for all the material we will need for backfill at Res. Site.
>
> Reworked access Rd where test failed from Sta. 0+50 to 1+50. Dug out about 3' & backfilled with dryer material from top of access Rd. Re-test went to 96. Tests are still taking a longer time than is required. Was supposed to get test at 3:00pm. Soils man didn't show up until 3:35pm
>
> *Mick Jordan*
>
> *[Stamp: Data used are fictitious for illustration only]*
>
> Friday 7 February 1986
>
> Started making fill on access Rd by pushing material from Res. Site to access Rd fill with D-9 & 46-A.
> 46A went down for repair at 10:00. Called Chuck for another Cat, won't get one till noon Monday. Makes fill go pretty slow, but tests are coming out good. Mixing material with Cats & pushing seems to dry out the dirt a quite a bit. Should have a total of 3 Cats. Should put 1 Cat up making 1:1 Cut in back side of Res. and stockpiling dirt for backfill in back of Res. where the get it Ripped Cat area so it can dry out over the weekend
>
> *Mick Jordan*
>
> Monday 10 February to Thursday 13 February 1986
> No work — Rain
>
> *Mick Jordan*

FIGURE 9.7 Daily construction diary.

B&H CONSTRUCTION

GENERAL
PROJECT STATUS REPORT

PROJECT: Arroyo Wastewater Treatment Plant Addition
OWNER: City of Casa Grande
PROJECT NO.: 07-01-28675
CONTRACT NO.: 88-0324
PROJECT MANAGER: Ray E. Barnes
DATE: 10 June 1988

Amount of Original Contract . . . . $	6,512,340.00
Approved Change Orders to Date	6,203.00
Anticipated Over-run or (Under-run) in Uncompleted Work	2,562.00
Actual Over-run or (Under-run) in Completed Work	
Rental Revenue from Contract Owner not included above	
Estimated Total Amount of Principal Contract . . . . $	6,521,105.00
Other Contract Work not Included in Principal Contract	
Total Estimated Contract Volume . . . . $	6,521,105.00

Contract Revenue to Date . . . . $	2,717,604.00	
Less: Contract Advances for Materials on Hand	(265,712.00)	
Contract Advances for Plant and Move-in	(105,842.00)	
Revenue Reduction for Uncompleted Work	(80,601.00)	
Other		
Total Amount of Work Completed to Date . . . . $		2,265,449.00

Uncompleted Contract Volume . . . . $	4,255,656.00
Percent Complete Based on Original Contract	34.79 %
Percent Complete Based on Total Estimated Contract	34.74 %
Time Allotted by Original Contract	462 Days
Extension of Contract Time	10 Days
Total Contract Time	472 Days
Contract Time Elapsed	145 Days
Percent of Original Contract Time Elapsed	31.39 %
Percent of Total Time, Including Extensions, Elapsed	30.72 %
Date Contract was Physically Completed—If Completed	
Expected Date of Physical Completion—If Not Complete	3 May 1989

Data used are fictitious for illustration only

Wiley-Fisk Form R-7

FIGURE 9.8 Monthly report of project time and cost status.

the accumulative totals of the various monthly progress payment documents.

For individual project cost monitoring, the form shown in Figure 9.9 can be of considerable benefit to many contractors. By not only tabulating the monthly accumulations of budget and actual cost, but by graphically plotting each amount on a time vs. cost chart, similar in format to the traditional S curve chart, a visual comparison can be made that clearly indicates not only the status of the contract at any given time, but will also show any change in a trend toward either a predicted savings or a cost overrun. By plotting a curve representing the amounts invoiced to the agency for monthly progress payments, an additional dimension is provided. To prepare and keep up such a chart, regular inputs are required by the project manager or project superintendent on all costs of labor, materials and equipment, rentals, subcontracts, overhead, profit, and so on. Such information must be made available on a regularly scheduled basis at the end of each billing cycle.

Requests for Services

From time to time, the contractor will require services that are provided by either the agency or the architect-engineer. These may be included in the contract, but must be requested on an as-needed basis by the contractor. Typically, such contracts contain language regarding the costs of resetting survey markers when they have been moved, damaged, or knocked out by the contractor's personnel. In addition, there are frequent disputes over the time of a request by the contractor vs. the time of its receiving the needed services, or the scope of those services. As a protection to itself, a contractor should make all such requests in writing so as to have documentation to support its position in case of a later dispute over the issue. A form designed for the documentation of a contractor's request for a survey is shown in Figure 9.9. Many forms will do the same job, however; this merely serves as an example of the method of documentation.

TEMPORARY FACILITIES

Temporary facilities generally expected of the contractor include all temporary power and construction lighting for the site, including a mandatory separation of power and lighting circuits. Such wiring is required to meet all local codes and ordinances. The water supply is often up to the contractor to

E. R. FISK CONSTRUCTION

REQUEST FOR SURVEY

Project: Green Valley Interceptor & Collector Sewers
Date: 16 June 1987
Title: Carson Interceptor Line
Item: Lateral 5-A

TO: LFM Consulting Engineers, Inc.
SUBJECT: Sewer Grades and cut sheets

Request surveyor for Monday 22 June 1987

Need the following staking done:

1. Restake alinement from Sta 154+56 to 155+05 which was knocked out as the result of highway accident previously reported

2. Sewer grades and cut sheets needed for Sta 144+25 to 149+37

Please contact Jorgensen & Sons, pipeline Supt for details

Data used are fictitious for illustration only

Please note latest acceptable completion date: 25 June 19 87.

Requested By: E. R. Fisk
Project Manager
Title: L&M Construction Co., Inc.
Contractor: General Contractors

Wiley-Fisk Form 9-3

FIGURE 9.9 Documentation of contractor's request for survey.

locate and make arrangements for, in addition to installing its own temporary pipelines, valves, pumping system, hydrants, storage tanks, and all other equipment necessary. Generally, contractors are prohibited from drawing water from fire hydrants for construction purposes. In some cases, where potable water is not available, separate bottled water shall be provided for all employees at the site. The obligation to provide a temporary water system includes the obligation to remove all temporary facilities at the close of construction. Adequate fire protection must also be provided, meeting all OSHA and local fire code regulations.

Field Offices

On many projects, it is the contractor's obligation to provide field offices for itself *and* the agency or architect-engineer personnel assigned to the site. Failure to meet the strict requirements of the contract documents could be interpreted as default, and a contractor is well advised to provide promptly such facilities and in strict compliance with the requirements of the contract documents.

Communications

As a part of a contractor's obligations to provide field offices, it is normally required to provide all field telephones in construction field offices. The contractor, under such contracts, must install a telephone in its own field office in addition to one in the agency's or architect-engineer's office. The contractor is further obligated to provide free use of such telephones to all authorized users, but is usually allowed to charge all message unit and long distance toll calls to the agency at the rates charged by the telephone company.

One of the biggest problems with telephones, and one that will get a contractor off on the wrong foot at the beginning of the job, is failure to install the agency's or architect-engineer's field office telephone on time—or, in some cases, at all. It is to a contractor's benefit to meet these rather modest requirements as early as possible and avoid the risk of developing a bitter adversarial relationship even before a job starts rolling.

Environmental Controls

Public works contracts typically contain requirements for contractor precautions and obligations on environmental issues. These include the provision and maintenance of sanitary facilities for all employees and the regular

TEMPORARY FACILITIES

collection of sanitary wastes, dust abatement measures, air pollution control measures, noise abatement provisions, rubbish control, disposal of hazardous wastes, and explosives and blasting. In addition, the preservation of cultural resources is covered under the National Historic Preservation Act of 1966, which includes provisions for the protection of potential historic, architectural, archaeological, or other cultural resources. Sufficient consideration should be made of such items in preparing a bid, as these provisions are subject to rigid enforcement in most jurisdictions.

Project Signs

A cost not often anticipated by contractors whose previous experience was limited to private work is that of an elaborate project sign. Typically, on

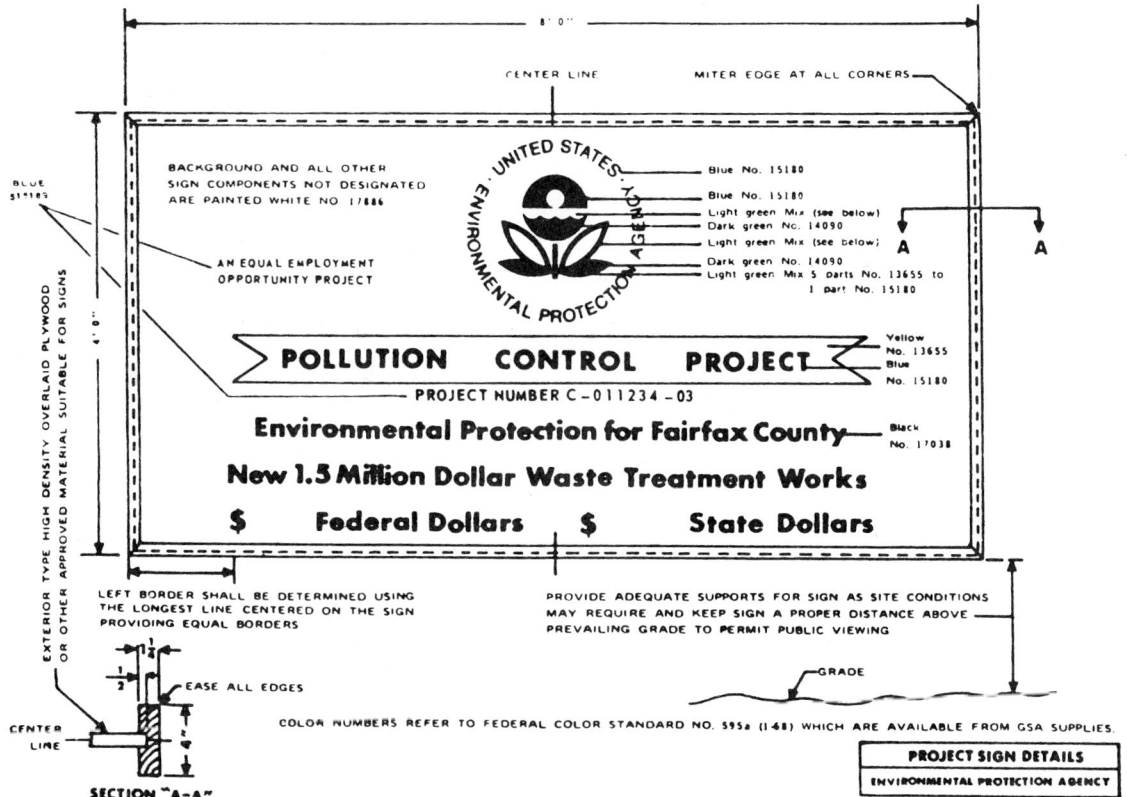

FIGURE 9.10 Project sign as required on EPA-funded projects.

federal and federally funded projects, a large, sometimes costly sign is required under the terms of the contract documents. Look for this provision; it is often buried in the front-end documents, rather than the technical section of the specifications.

A classic example is the sign required by the EPA for projects being constructed with federal funds under the Clean Water Grant Program, the requirements of which are shown in the specification page illustrated in Figure 9.10. Many such projects have been built by cities, counties, and special districts under a federally subsidized program. Failure to budget for a sign in this kind of project can contribute to cost overruns and an inevitable reduction in profitability.

SCHEDULING AND COORDINATION

Scheduling techniques such as CPM or other network-scheduling systems, and their associated computer-generated printouts have sometimes only served to compound the problem. Often, the contractor has stacks of computer printouts that it either does not want or does not know how to read. But somewhere in that stack of paper may well be the key to a contractor's success in prosecuting or defending against claims. The question of scheduling methods of operation and the impact of deviating from an anticipated schedule require careful analysis to determine the reasonableness of the originally planned schedule in conjunction with the planned methods of operation. The analysis of every schedule-related problem is unique, and every such analysis should include a review of the anticipated sequence and schedule, together with a review of the actual progress of the work. The review must include an analysis of any delay and impact caused by all parties to the progress of the project.

CPM schedules are especially helpful for claims and change orders arising from time-related problems such as delays, suspensions, or accelerations of the work. Contractors have historically resisted CPM scheduling, not because they find fault with its principles or theory, but because of the numerous administrative problems it can cause within their own organizations. First, it forces a contractor to study its operation in detail, with careful planning of all sequences. Within a very short period at the beginning of a job, the contractor must plan its construction methods for a job that may have taken two years to design and may well take three years to build. Second, as CPM is a logical sequence of events and details task duration, obvious errors can easily be spotted. This can become an embarrassment to the contractor. Finally, CPM becomes a commitment involving great detail, for all to see. However, done properly, it can be the contractor's greatest defense tool when justifying impact costs of agency-generated change orders or unforeseen site conditions or delays.

More often than not, a CPM schedule will aid the contractor when it must file claims. A good CPM usually favors the contractor and is worth a thousand pictures and ten thousand letters. CPM schedules are recognized by owners, architect-engineers, contracting officers, boards of contract appeals, and the courts.

Schedule Submittals

Most public works contracts require the contractor to submit schedules to the architect-engineer for review and approval prior to the beginning of the work. It should be borne in mind that this is a requirement that the contractor cannot afford to overlook. Such schedules should be carefully prepared to benefit the contractor and its operation. In addition, the contractor is required to submit updated schedules at predetermined periods, and in some cases, at any time when the agency or the architect-engineer has reason to believe that the contractor may not finish the project on schedule. Whenever the contractor is significantly delayed in the work, the agency may require a revised schedule to be prepared and submitted to the agency or its architect-engineer for approval, showing how the contractor plans to revise its construction operations so as to complete the work on time.

Form of the Schedule

Whenever the Contract Documents do not specifically require a CPM network or another form of schedule, the contractor has the right to determine the form of the schedule. Although a bar chart may not be the best way to attempt to schedule and control a construction project, it is often within the contractor's rights to submit a schedule in the format of its choice.

Whenever CPM scheduling is called for, however, the agency is within its rights to demand that the contractor provide and use such scheduling methods to perform the work regardless of the cost to the contractor for providing such services.

When estimating scheduling requirements for the bid, look carefully at the scheduling requirements to determine the size and frequency of updating the network diagram as well as the computer printout format that must be provided, along with the number of data "sorts" that must be provided and the frequency of submitting such printouts. It is not unusual to find requirements that the contractor update the computer printouts at weekly or bi-weekly intervals, and that the data be furnished in as many as three different sorts or arrangements of all project activities.

CPM and Cost Reporting

CPM can be used by the contractor to aid the project manager in the cost aspects of project control. This aid comes in various forms: as predictions for the need for cash, as reports on money spent vs. work accomplished, as reports verifying subcontractor bills, and as statements of money due to the contractor. Examples of an application of cost-loaded CPM schedules can be found in the AGC publication entitled *The Use of CPM in Construction* (Associated General Contractors of America, Washington, D.C.).

Occasionally, the agency specifications require the contractor to supply cost-loaded CPM schedules for use in estimating construction progress payments. This is always a sore spot with contractors as they traditionally feel that such cost data are privileged information. Whether it is or is not does not change the fact that, under such specifications, the contractor certainly must produce some kind of cost-loaded network. For convenience in estimating monthly progress payments, it would be best to set forth schedule activities that correspond to the pay line items on the bid sheets or the schedule of values. The conversion of a contractor's work activities into cost values corresponding to 20 or 30 pay line items is a major operation. It is far simpler to program a PC computer for handling a short progress payment program. However, if an agency cannot be talked out of requiring cost values for all work activities, the contractor is obligated under the contract to provide the information called for in the contract documents or it will be in default.

CONTRACT TIME

The contract time may be stated in either working days or calendar days, or as an alternative, by stating a calendar date for completion. Calendar-day contracts, although easier to compute, should be looked upon as having been planned by the owner or architect-engineer to provide the requisite number of working days plus weekends and holidays to arrive at the total time. When negotiating time for a change order on a calendar-day contract, be certain to add in the number of weekends and holidays to arrive at the total number of days required.

"Time is of the Essence"

These key words, "time is of the essence," are important to every contractor on a job that calls for liquidated damages. Failure of the contract documents to contain this phase renders the contract completion date a *target date,* not an absolute date in the view of the court. See the discussion of liquidated damages in Chapter 12 for further discussion of this subject.

Stopping or Suspension of the Work by the Architect-Engineer

Whenever the architect-engineer stops or suspends all or part of the work for any cause but nonconforming work or a serious safety hazard, the contractor has reasonable cause to believe that recovery for the cost and time of the delay may be possible. Such recovery may include not only direct costs, but impact costs on other activities and possible unabsorbed or extended home office overhead. The probability of such recovery, however, is heavily dependent upon solid evidence in the form of sufficient documentation. Those contractors utilizing a network scheduling system, such as CPM, stand the best chance of proving the extent of such costs.

Delays resulting from the rejection of nonconforming work, however, are not only nonexcusable, they may actually cost a contractor in terms of liquidated damages if they contribute to a project overrun beyond the contractual completion date.

Similarly, agency or architect-engineer suspensions of work in hazardous locations due to the failure of the contractor to provide for proper safety procedures are a risk that must be borne solely by the contractor.

DIFFERING SITE CONDITIONS

Several of the provisions of the General Conditions deserve special discussion because of their importance to the contractor as a result of their possible varying interpretations. One of the most misunderstood of all contract provisions and the one that is frequently the cause of large contractor claims for additional work and change orders is the matter of *differing site conditions.*

CHAPTER 9   PROJECT ADMINISTRATION

Unforeseen Underground Conditions

Subsurface and latent physical conditions at the site present a special problem. If they differ significantly from what is covered in the contract documents, the contractor may well be entitled to additional payment for any extra work involved. When this happens, it usually comes as a great surprise to the public agency. The architect-engineer will normally be extremely careful to give no assurance to a bidder as to the accuracy of any subsurface exploration, even when a special soils consultant was employed. Failure of the public agency or architect-engineer to advise the contractor of any available data regarding subsurface conditions may not only entitle the contractor to additional payment, but may possibly be the cause of a significant delay claim when underground conditions are discovered that are quite different than shown on the plans.

The public agency or architect-engineer can no longer say that the responsibility for all such conditions and the delays that accompany them will always be that of the contractor. Recent court decisions have relieved the contractor of such responsibility even where the wording of the contract documents stated that the contractor must familiarize itself with all conditions at the site that might affect the performance of the work. In California public works contracts this is carried one step further, as the code specifically requires the public agency to assume all responsibility for the relocation or removal of existing utilities or pipelines that are discovered in an area the contractor must excavate.

In fairness to the contractor, to avoid the risk of blame for causing delays on the job, and as a means of reducing its exposure to claims, the architect-engineer or public agency must make all data that were used in the design available to the contractor.

The present-day approach by those public agencies endeavoring to be fair in their contracting policies is to pattern their provisions for differing site conditions along the lines of those developed by the federal government, which are widely accepted as a reasonable means of handling unforeseen problems.

The provisions of the federal contract forms call for the making of adjustments in time and/or price when unknown subsurface or latent conditions at the site are encountered. The purpose is to have the government accept certain risks and thus reduce large contingency amounts in bids to cover such unknown conditions. The government includes provisions in its construction contracts that will grant a price increase and/or time extension to a contractor who has encountered subsurface or latent conditions.

CHANGE-ORDER ADMINISTRATION

The typical "Differing Site Conditions" clause in federal contracts reads as follows:

(a) The Contractor shall promptly, and before such conditions are disturbed, notify the Contracting Officer in writing of (1) Subsurface or latent conditions at the site differing materially from those indicated in this Contract, or (2) Unknown physical conditions at the site, of an unusual nature, differing materially from those ordinarily encountered and generally recognized as inherent in work of the character provided for in this Contract. The Contracting Officer shall promptly investigate the conditions, and if he finds that such conditions do materially so differ and cause an increase or decrease in the Contractor's cost of, or the time required for, performance of any part of the work under this Contract, whether or not changed as a result of such conditions, an equitable adjustment shall be made and the Contract modified in writing accordingly.

(b) No claim of the Contractor under this clause shall be allowed unless the Contractor has given the notice required in (a), above; provided, however, the time prescribed therefore may be extended by the Government.

(c) No claim by the Contractor for an equitable adjustment hereunder shall be allowed if asserted after final payment under this Contract.

It is common to refer to such differing site conditions provisions as being a "Type 1" or a "Type 2" Differing Site Condition, based upon the definition contained in Paragraphs (a)(1) and (a)(2), above, and summarized below (see also Chapter 11 on "Differing Site Conditions").

Type 1 Differing Site Condition
Subsurface or latent physical conditions at the site which differ materially from that indicated in the contract documents.

Type 2 Differing Site Condition
Unknown unusual physical conditions at the site which differ materially from the conditions ordinarily encountered and recognized as inherent in this type of work.

CHANGE-ORDER ADMINISTRATION

A change order is the formal document that alters some condition of the Contract Documents. The change order may alter the contract price, schedule of payments, completion date, or the plans and specifications. It specifies the agreed upon change to the contract and should include the following information:

1. Identification of the change order.
2. Description of the change.
3. Reason for change.
4. Change in contract price.
5. Change in unit prices (if applicable).
6. Change to contract time.
7. Statement that secondary impacts are included or excluded.
8. Approvals by *both* the agency and contractor.

The agency may require the architect-engineer's signature for its own purposes, but all that is required is the agency's and the contractor's signature. Both *must* sign a change order.

A change order is a written agreement between the agency and the contractor to perform an addition, deletion, or revision of work and/or the time of completion within the limits of the scope of the construction contract after it has been executed (see Figure 9.11). It is a specific type of contract modification that does not and cannot go beyond the general scope of the existing contract.

The change order generally originates as a claim or recommendation for a change from the contractor (see Figure 9.12) or as a request for proposal from the public agency, seeking a change to the existing Contract Documents. The change order is necessary to increase or decrease the contract cost or work, to interrupt or terminate the project, to revise the completion date, to alter the design, or, in general, to implement any deviation from the original contract terms and conditions through a bilateral agreement.

The following are common categories of conditions that generally give rise to a need for a contract change:

1. Differing site conditions.
2. Errors and omissions in the plans and specs.
3. Changes instituted by regulatory agencies.
4. Design changes.
5. Quantity overruns/underruns beyond limits.
6. Factors affecting time of completion.
7. Adjustments to time or cost.

CHANGE ORDER

PROJECT TITLE: Inglewood Water Project - Unit III Water Treatment Plant
PROJECT NO. 415-26   CONTRACT NO. 88-4321   CONTRACT DATE 5-25-87
CONTRACTOR: International Constructors, Inc.

The following changes are hereby made to the Contract Documents:

Construction of basin drainage system; and reset two pump bearing plates. In accordance with revised drawing S-17209 Rev. 5, dated 26 August 1988.

Justification:
Unforseen soil conditions

Data used are fictitious for illustration only

CHANGE TO CONTRACT PRICE

Original Contract Price: $ 6,231,053.00

Current contract price, as adjusted by previous change orders: $ 6,257,760.85

The Contract Price due to this Change Order will be [increased] [decreased] by $ 4,342.00

The new Contract Price due to this Change Order will be: $ 3,262,102.85

CHANGE TO CONTRACT TIME

The Contract Time will be [increased] [decreased] by –14– calendar days.

The date for completion of all work under the contract will be 31 August 1988

Approvals Required:

To be effective, this order must be approved by the Owner if it changes the scope or objective of the project, or as may otherwise be required under the terms of the Supplementary General Conditions of the Contract.

Requested by *E. J. King*   Proj Mgr, LFM Consulting Engrs, Inc.   date 26 Aug 1988
Recommended by *[signature]*   LFM Consulting Engineers, Inc.   date 26 Aug 1988
Ordered by *[signature]*   City of Inglewood, CA   date 31 Aug 1988
Accepted by *T. V. Capehart*   International Constructors, Inc.   date 6 Sep 1988

Wiley-Fisk Form 15-3

FIGURE 9.11 Typical change-order form.

CHANGE ORDER REQUEST

E. R. FISK CONSTRUCTION

Project Title: RTD Bus Maintenance Facility
Project No.: 87-0243   Contract No.: _____   Contract Date: 9-22-87

Proposed By: G. B. Taylor, Supt   Date: 7-26-88
                (Name)
Submitted By: R. E. Barnes, Project Manager   Date: 8-01-88
                (Name)

Actual job conditions in area of proposed change:
All masonry work completed; rolling door installed & operable; Electrical service not yet installed

Change order justification:
Area served by rolling door has no other access or openings. Design of rollup door installation calls for operating controls to be installed inside the room. Door cannot be operated from outside.

Contractor authorized to proceed with this change ☐ YES ☐ NO on _____

Other contracts involved are as follows (List Contracts by No.): none   Is Dwg. Req.? ☐ NO ☒ YES E-14
                                                                                                    (Sheet No.)

Description of Work to be Performed:
Drill masonry wall to allow conduit installation from maintenance area into the security room and install electricall wiring to allow electrical motor operation of the door to the secure area from the adjacent maintenance room. Install a key-operated switch for actuating the rolling door to the secure area. Power from circuit J-3.

Data used are fictitious for illustration only

Estimated effect on Cost: $1080.00   Time: 3 days

R.E. Barnes   8-01-88
R.E. Barnes, Proj. Mgr.   (Date)

Wiley-Fisk Form 15-2

FIGURE 9.12 Contractor's request for a change order.

CHANGE-ORDER ADMINISTRATION

It is possible that some changes may fall outside these general categories; however, others commonly have characteristics similar to the categories just outlined. Therefore, they can be related to the reasoning process developed in one or more of the seven mentioned categories.

It is extremely important to maintain an up-to-date log (see Figure 9.13) and change-order control sheet (see Figure 9.14) for all change orders and the action taken, including the dates and current status.

Factors the Agency Will Take into Consideration

Before a public agency will consider a contractor's request for a change order justifiable, it will ask several critical questions.

1. Was the cause of any delay beyond the contractor's control? Did the contractor take normal precautions?
2. Was the contractor ready and able to work?
3. Did the contractor submit a detailed schedule projecting completion within the allotted time?
4. Was the schedule updated regularly?
5. Did the updated schedule justify a time extension?
6. Did the schedule contain a critical path analysis or equivalent?
7. Has the contractor maintained sufficient forces in those operations along the critical path where needed to meet the target dates?
8. How have causes other than normal weather beyond the control and without the fault or negligence of the contractor affected target dates along the critical path?
9. Has the contractor proven "unusually severe weather" with such information as climatological data, the probability of severe storms, or flood-depth data?
10. Did the weather phenomenon actually delay operations along the critical path or secondary operations?
11. Was the contractor shut down for other reasons?

Change Orders and Extra Work in Subcontracts

Virtually every construction project at one time or another is subject to a change order or an extra. Most prime contracts recognize that fact and contain provisions defining extras and change orders: how they are made and

• 201 •

CHANGE ORDER LOG

E. R. FISK CONSTRUCTION

JOB NO. 0700.001 PROJECT 15 MGD Plant Addition OWNER Hidden Valley WTP

C.O. NO.	DESCRIPTION	COST	CREDIT	CUMMULATIVE AMT. TO DATE	EXTENSION OF TIME	ADJUSTED TOTAL CONTRACT TIME
1.	Misc. items	$ 10,120		$ 10,120	none	540 da
2.	Misc. items	40,380		50,500	none	540 da
3.	Cons. Time Extension	–		50,500	+180 da	720 da
4.	Time Ext. & Sludge Removal	186,649		237,149	+ 60 da	780 da
5.	Misc. Items	58,996		296,145	+ 70 da	850 da
6.	Delete access road paving		309,705	-13,560	- 60 da	790 da
7.	Const. Temp. Filter Eff. Line	31,250		17,690	none	790 da

FIGURE 9.13 Example of a change-order log.

CHANGE ORDER CONTROL SHEET

E. R. FISK CONSTRUCTION

Project Title: Pumping Station D-3
Project No.: 88-01239 Contract No.: W07-01-7658 Contract Date: 14 March 1988

Change Order Number	Number	Phase of Work (Gen. Mech. Elec.)	Date Initiated	Proposal Received	Approved	Order Issued	Cancelled	Amount of Change Order Add	Deduct	Work Delayed From	To	Date Work Started	Completed	Extensions Approved	Recommended	Posted on Pay Sheet
1	4A	Masonry	4/22	5/9	5/16	5/16		0	0	—	—	5/19	6/3			
2	6B	Gen.	6/13	6/14	6/14	6/14		1,295		11/16	11/25	4/20	6/29	6	P.D.	✓
3	15C	Mech.	8/3	8/8	8/10	8/12			<695>	—	—	8/15				

Wiley-Fisk Form 15-4

Data used are fictitious for illustration only

FIGURE 9.14 Documentation of change-order status.

how the contractor will be paid for them. If a particular approval procedure is to be used, the subcontracts must reflect that fact. If the contractor will not receive payment for a change order or extra work until the funding agency releases the funds, the subcontracts should reflect that information as well. The contractor may find it necessary in some contracts to make payment to its subcontractors before it gets paid for the subcontractors' work. From the subcontractor's viewpoint, this is the most desirable approach, as the subcontractors resent being used as the "bankers" for the contractor's project.

Field Order or Work Directive Change

A field order, or Work Directive Change as it is called by the EJCDC in the standard documents of the Joint Documents Committee, is a unilateral change and was developed for use in situations involving changes in the work that, if not processed expeditiously, might delay the project. Such changes are often initiated in the field and may affect the contract price or the contract time. A Work Directive Change is not a change order, but only a directive to proceed with the work that may be included in a subsequent change order.

Under EJCDC standards, a Field Order may only be used for supplemental instructions and minor changes not involving a change in either the contract price or the contract time.

JOB CHANGES AND PERFORMANCE BONDS

As a contractor's bonding capacity is a valuable asset, a public works contractor must protect it at all costs. Particular care must be taken to ensure that the contractor works in conjunction with its bonding agent and surety as the job progresses. For example, changes in the work on a particular project may increase the risks of default on the part of the contractor. Under general suretyship law, such changes in the contractor's obligations incurred after the commencement of a contract and without the surety's consent may result in the surety being released from all obligations to the agency. Many performance bonds get around such problems by providing that the surety waive notice of all changes or alterations in the contract. Some performance bonds issued for government construction require that the surety be advised of all changes that correspond to a cost increase about a certain percentage of the original contract price.

If a bond on a particular project were to have such a provision, the contractor should develop an administrative system whereby the surety's notice requirements would be met. It is suggested that under such circumstances, the surety receive copies of all change orders as they arise.

When the surety bond is written in such a way that changes are authorized without the surety's prior consent, the changes must be within the general scope of the contract. Any changes beyond the scope, and not within the reasonable contemplation of the parties involved at the time of entering into a contract, could be deemed radical or major changes. Without the surety's prior consent to changes that are outside the originally contemplated scope of the contract, such changes may result in its release under a performance or payment bond, even though it contains a waiver-of-notice provision. Although the courts have been reluctant to characterize changes in work as *radical* or *major*, the whole problem can be avoided altogether by implementing administrative procedures that keep the surety advised of all changes on the project. Even if the surety does not consent to the changes, the mere fact that it has been put on notice of such changes will shift the burden to the surety to voice its objection to such changes.

LABOR AND MATERIAL RELEASES

It is good practice for a contractor to insist that subcontractors present labor and material releases for the work that they have performed, as well as labor and material releases for their suppliers and subcontractors as a condition of receiving payment from the general contractor. Such releases may be conditional or unconditional, but are generally only effective on the releasing party for the amount of money due and stated in the release itself.

A contractor must obtain such releases in order to be assured that money being paid out to subcontractors is, in turn, being paid to the proper parties. If proper payments are not being made by the subcontractors, the general contractor may ultimately be liable for such payments through its bonds, or funds due to the general contractor may be withheld by the public agency by way of a lien against the construction funds (Stop Notice).

Although a contractor may believe that the release of a remote subsubcontractor or material supplier will release the general contractor as well as its bond, it may not necessarily be so unless the release document contains specific language to that effect.

In the California case of *Powers Regulator Company v. Seaboard Surety Company of New York,* [204 Cal.App.2d 338 (1962)] a remote subcontractor on a public works project brought action against the contractor's labor and material bond (Payment Bond) even though it had released the job in writing from all claims for labor and materials. The subcontractor recovered on the bond even though the subcontractor would not have been paid by the contractor without the written release. As the release had been written for the benefit of the contractor and not for the benefit of the surety, the subcontractor had not released any right to recover from the bonding company. Consequently, any release should contain language that clearly states that the release is for the benefit of the sureties as well as the agency and the contractor.

As an alternative to obtaining releases, many contractors use joint checks to pay subcontractors and their suppliers. The endorsement signatures on the canceled check become proof that the parties have been paid. However, as just indicated, the canceled checks may be evidence of payment, but will not necessarily release a surety of its responsibility.

PAYMENT

There are several items to be considered in obtaining payment for work done on public works projects. The subject of payments includes consideration of the following categories:

1. Mobilization or "move-on" costs.
2. Progress (partial) payments during construction.
3. Work beyond specified contract limits.[3]
4. Materials delivered but not yet installed.
5. Retainage from earned funds.
6. Liquidated damages for late completion.
7. Final progress payment.
8. Withholding for uncompleted punch list items.
9. Release of retainage.

[3] Note that the term used says beyond contract limits, *not* beyond contract *scope*. It is a violation of the public bidding laws for a public agency to attempt to change the scope of a project without a separate advertisement for bids for that portion of the project beyond the original scope.

MATERIALS AND METHODS

Often, an architect-engineer will require that a contractor furnish *manufacturer's* certificates for certain materials. It is important that the contractor realize that *supplier certificates* will probably be rejected, and that the contractor runs the risk of added cost to belatedly obtain manufacturer certificates or may find the product in question rejected. Advise the supplier in advance, in writing, that manufacturer's certificates are required.

Handling Rejected Work

What should be done when the inspector rejects your work? The best course of action is to follow the instructions of the agency's or architect-engineer's inspector and take whatever measures that the inspector directs. This removes the possibility that the contractor can be considered as a contributor to any delays and associated impact costs. This is not to say that the contractor should absorb any unwarranted extra costs, if unjustified under the terms of the contract. The contractor is still left with the ability to file a claim for recovery after assembling the necessary documentation. One of the most important considerations is to document all such actions carefully and accurately.

Nonconforming Work

It is uncommon these days to be required to tear out nonconforming construction and rebuild, as the resulting delays often cost the agency more than the loss of the quality product. However, it is common for the parties to negotiate a monetary settlement to compensate the agency for receiving a product that is worth less than the one for which it contracted.

A contractor should be prepared to settle for less when work is rejected as nonconforming. It is still possible, of course, to be required to remove a nonconforming item, but many such incidents can be settled through negotiation instead. Whatever you do, be certain to settle the matter while it is simply an issue between two people: you and the agency's or architect-engineer's representative. The minute either side shows up with an attorney, the cost rises to cover the added legal cost. The result is that even a negotiated compromise settlement will be more costly than a total loss would have been when the stakes were lower.

CHAPTER 9 PROJECT ADMINISTRATION

Substitution of Products

One of the most common occurrences during construction is the constant search by the contractor to obtain substitute products that are less costly than those actually called for in the specifications and to offer them to the public agency or architect-engineer as substitutes under the "or-equal" provisions of the specifications. Great care must be exercised by the public agency or architect-engineer in the approval of new, substitute products as well as in the application of some established ones. The architect-engineer of record has the duty to see that the products furnished in compliance with the drawings and specifications are actually suitable for the particular uses intended. Reliance on a producer's sales literature is hazardous at best. There have been several court decisions in which the architect-engineer of record has been held liable for failure to have a new material tested, or an established item tested for a new application prior to approving it.

A public agency or architect-engineer may even require that the manufacturer of such new products furnish guarantees that extend beyond the usual time. The refusal of a producer to provide such guarantees may be considered sufficient reason for rejecting the product. In any case, the authority for accepting a product offered as an "or-equal" substitute is reserved to the architect-engineer. This can be challenged in court, however, if the or-equal product offered is deemed to be equivalent in quality and utility and yet is rejected by the agency or the architect-engineer. The or-equal product need *not* be identical to the specified product.

Value Engineering During Construction

CHAPTER • 10 •

- Definition of Value Engineering
- Fundamentals of Value Engineering (VE)
- Value Engineering by the Contractor
- Methodology in Generating VE Proposals

CHAPTER 10 VALUE ENGINEERING DURING CONSTRUCTION

Value engineering: What is it and what part does the contractor play in its application? The latter question will be answered in this chapter. The first question, however, deserves special consideration.

Some feel that value engineering is not a proper role for a contractor. Nevertheless, a construction project may involve construction methods, which the contractor has valuable knowledge to contribute to the architect-engineer decision maker. Under many federal contracts, it is an integral part of the contract requirements for the general contractor to propose value engineering ideas *during the construction phase*. There is no way that a contractor can avoid the issue when it is a requirement of the Contract Documents; thus, it is highly desirable for the contractor to know what value engineering is and how to work with value engineering proposals. In this chapter, the contractor will be introduced to the subject in sufficient detail to understand what it is all about and where it fits into the contractor's responsibility area.

A great deal of lip service has been paid to the concept of value engineering in recent years, and older engineers remember that years ago it was called engineering economics and every engineer was expected to apply its principles to all design projects. Now, it is implied by the proponents of this approach that value engineering is a new, previously undeveloped branch of the construction industry. There are entire engineering organizations devoted solely to the promotion of value engineering and at least one federal agency that provides grant funds for large construction projects and now requires its use under separate contract during the design phase of every grant-funded project it supervises.

The concept of providing value engineering incentives to the construction contractor during the construction phase of a project, however, is a relatively new application, and a deserving one. Often, the selection of materials and construction methods, sequences, and techniques by a public agency or architect-engineer is dictated by an evaluation of average market conditions, a superficial knowledge of construction equipment and methods, and anticipated contractor methods. When a contract for construction is awarded to a specific contractor, its particular skills, equipment, material sources, methods, techniques, and knowledge of the local trade area and labor market can often be used to beneficially reduce the cost of a portion of a proposed project without compromising the design concepts involved. The final judgment of the acceptability of such suggestions by the contractor would be up to the architect-engineer and the public agency.

DEFINITION OF VALUE ENGINEERING

The first question usually asked is, "What is value engineering anyway?" Value engineering is a systematic evaluation of a project design to obtain the most value for every dollar of cost. By carefully investigating costs, availability of materials, construction methods, shipping costs or physical limitations, planning and organizing, cost/benefit values, and similar cost-influencing items, a reduction in the overall cost of a project can often be realized.

The entire value engineering effort is aimed at a careful analysis of each function and the elimination or modification of anything that adds to the project cost without adding to its functional capabilities. Not only are first costs to be considered, but the later in-place costs of operation, maintenance, life, replacement, and similar characteristics must be taken into account as well. Thus, although the term is relatively new, value engineering is simply a systematic application of engineering economics as taught in every engineering course long before anyone ever thought up a catchy name for it.

During the construction phase, the value engineer is the contractor, whose experience lies in construction methods, techniques, and costs. The contractor can offer the benefit of its construction experience and current knowledge of the marketplace and labor force that the designer does not possess. It is here that the greatest cost benefit can be realized with a minimum of conflict with the designer. Often on federally funded construction contracts, a value engineering incentive clause may be provided in which the government will allow the contractor to retain 50 percent of any cost savings realized in any value engineering proposals submitted by the contractor and accepted and implemented by the engineer. Similar value engineering incentive programs are mandated by law in California and a few other states.

FUNDAMENTALS OF VALUE ENGINEERING (VE)

Function

In value engineering, *function* is defined as the specific purpose or use intended for something. It describes what must be achieved. For value engineering studies, this "function" is usually described in the simplest form possible, usually in only two words, a verb and a noun. "Support weight," "prevent corrosion," and "conduct current" are typical expressions of function.

CHAPTER 10 VALUE ENGINEERING DURING CONSTRUCTION

Worth

Worth refers to the least cost required to provide the *functions* that are required by the user of the finished project. Worth is established by comparison, such as comparing it with the cost of its functional equivalent. The worth of an item is not affected by the possibility of failure under the value engineering concept embraced by the federal government. Thus, if a bolt supporting a key joint in a large roof truss fails, the entire roof of a structure may possibly fail. Nevertheless, the worth of the bolt is the lowest cost necessary to provide a reliable fastening.

Cost

Cost is the total amount of money required to obtain and use the functions that have been specified. For the seller, this is the total of their costs in connection with their product. For the public agency, the total cost of ownership includes not only the purchase price of the product, but also the cost of paperwork and of including the product in the inventory, operating it, and providing support in the form of maintenance and utility services for its total usable life. The cost of ownership may also include a proportional share of expenditures for development, engineering, testing, spare parts, and various items of overhead expense.

Value

Value is the relationship of *worth* to *cost* as realized by the public agency, based upon its needs and resources in any given situation. The ratio of worth to cost is the principal measure of value. Thus, a "value equation" may be used to arrive at a *Value Index* as follows:

$$\text{Value Index} = \frac{\text{Worth}}{\text{Cost}} = \frac{\text{Utility}}{\text{Cost}}$$

The value may be increased by doing any of the following:

1 Improving the utility of something with no cost change.
2 Retaining same utility for less cost.
3 Combining improved utility with less cost.

Optimum value is obtained when all utility criteria are met at the lowest overall cost. Although worth and cost can be expressed in dollars, value is a dimensionless expression showing the relationship of the other two.

The Philosophy of Value

If something does not do what it is intended to do, no amount of cost reduction will improve its value. Any "cost reduction" action that sacrifices the needed utility of something actually reduces its value to the agency. On the other hand, costs incurred to increase the functional capacity of something beyond that which is needed amounts to "gilding of the lily" and provides little actual value to the public agency. Therefore, anything less than the necessary functional capacity is unacceptable; anything more is unnecessary and wasteful.

Types of Value Engineering Recommendations

Within the Defense Department and some other federal agencies, there are two types of recommendations that are the result of a value engineering effort:

Value Engineering Proposal (VEP)   A value engineering recommendation that originates from within the government agency itself, or one that was originated by the contractor and may be implemented by unilateral action. A VEP can only relate to changes that are within the terms of the Contract Documents and *thus would not require a change order to implement*.

Value Engineering Change Proposal (VECP)   A value engineering recommendation by a contractor that requires the public agency's approval and that, if accepted, *requires the execution of a change order*. This would apply to any proposed change requiring a change in the contract, the specifications, the scope of the work, or similar limits previously established by the Contract Documents.

VALUE ENGINEERING BY THE CONTRACTOR

Value engineering is a basic approach that takes nothing for granted and challenges everything on a project, including the necessity for the existence of a product or project for that matter. Prior to starting actual construction, the contractor should carefully consider the methods and equipment that may be used to construct the project. Requirements that increase the cost without producing equivalent benefits should be eliminated. One desirable characteristic of a successful contractor from the standpoint of value engineering is a degree of dissatisfaction over the plans and methods under consideration by the architect-engineer for constructing a project (a characteristic not always appreciated by either the public agency or the architect-engineer). However, complacency by members of the construction industry will not develop new equipment, new methods, or new construction planning, all of which are desirable for providing continuing improvements in the construction industry at lower costs. A contractor who does not keep informed on new equipment and methods will soon discover that its competitors are underbidding it.

Improvements in the methods of construction, long the sole domain of the contractor, can result in significant savings in the cost of a project. This type of cost saving, if implemented after the award of a contract, is seldom, if ever, shared with the public agency. However, such cost-reducing considerations are an integral part of the competitive bidding system. Thus, the public agency benefits in lower bid costs. As an example, an estimator for a contracting firm prepared a bid for a project. When the bids were opened, it was discovered that its bid was so low that the other members of the firm feared that a serious error had been made in preparing the bid. The estimator was called in and asked if he thought that he could actually construct the project for the estimated cost. The estimator replied that he could if permitted to adopt the construction methods assumed in estimating the cost. The firm agreed; he was placed in charge of the construction project and successfully completed the work with a satisfactory profit to the contractor. At the same time, the agency benefited by receiving its project at a low cost.

Other suggestions for possible reductions in construction costs by the contractor include, but are by no means limited to, the following:

1. Study the project before bidding and determine the effect of
   (a) Topography.
   (b) Geology.
   (c) Climate.
   (d) Sources of materials.
   (e) Access to the project site.

(f) Storage facilities for materials and equipment.
(g) Labor supply.
(h) Local services.

2. The use of substitute construction equipment that has greater capacities, higher efficiencies, higher speeds, more maneuverability, and lower operating costs.
3. Payment of a bonus to key personnel for production in excess of a specified rate.
4. The use of radios as a means of communication between headquarters office and key personnel on projects in large or remote areas.
5. The practice of holding periodic conferences with key personnel to discuss plans, procedures, and results. Such conferences should produce better morale among the staff members and should result in better coordination among the various operations.
6. The adoption of realistic safety practices on a project as a means of reducing accidents and lost time.
7. Consider the desirability of subcontracting specialized operations to other contractors that can do the work more economically than the general contractor.
8. Consider the desirability of improving shop and servicing facilities for better maintenance of construction equipment.

METHODOLOGY IN GENERATING VE PROPOSALS

A task that is accomplished in a planned and systematic manner is much more likely to be successful than one that is unplanned and relies on undisciplined ingenuity. Most successful value engineering organizations follow a "scientific method" to assure a planned, purposeful approach. This procedure is called a *VE Job Plan*. It is set up as a group action because it is unlikely that a successful value engineering proposal will be the product of a single individual. The group plan produces benefits that the efforts of one or two individuals can seldom match. Among the principal benefits are

1. More talent is directly applied to the problem.
2. The scope and depth of the effort are increased.

3 More efficient use is made of the available time because problem areas are most readily resolved through direct communication.
4 Team participation provides productive training for those not previously exposed to formal VE training.

Several versions of a VE Job Plan can be found in current publications. Some texts list five phases, some six, and some refer to even more. However, the number of phases is less important than the systematic approach involved. As an example, Figure 10.1 illustrates a five-phase job plan.

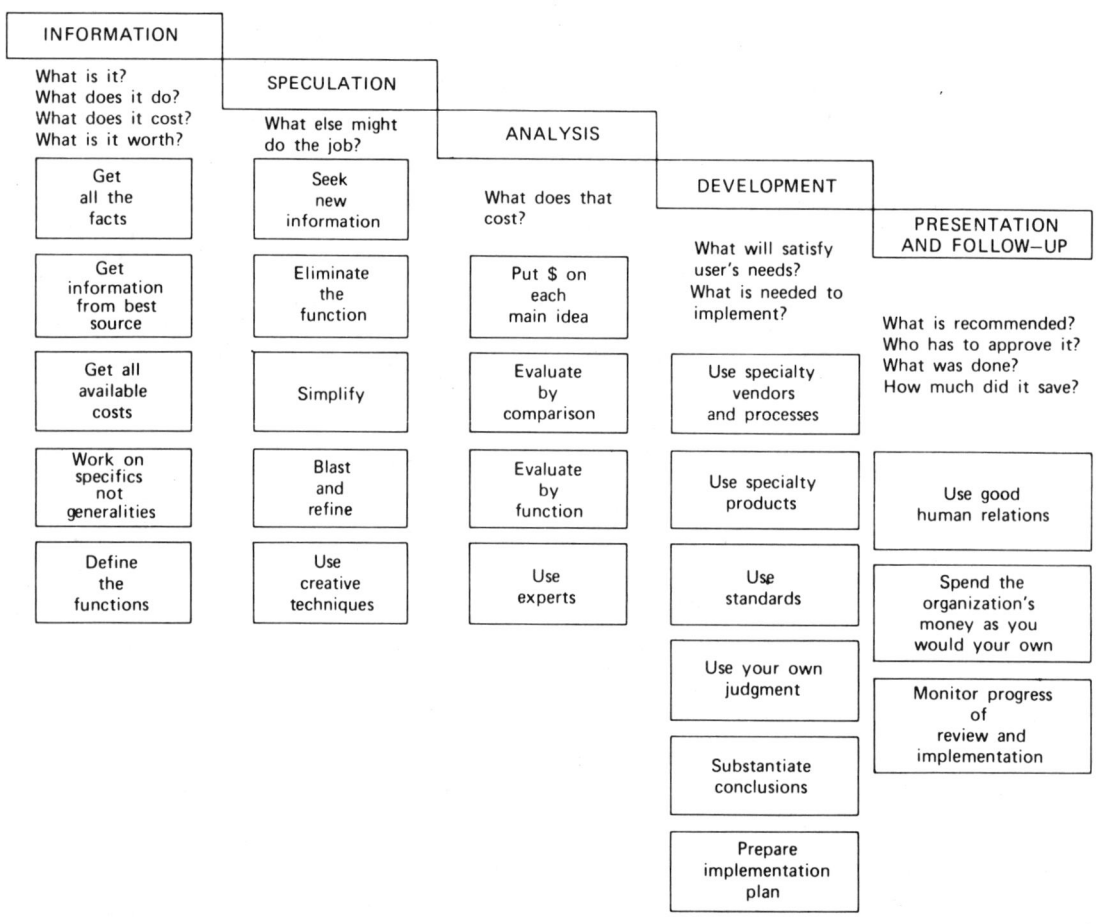

FIGURE 10.1 Five-phase VE Job Plan chart.

Although the illustration may suggest otherwise, there are actually no sharp lines of distinction between the phases. In practice, they tend to overlap to varying degrees.

An effective value engineering effort must include all phases of the Job Plan. However, the proper share of attention given to each phase may differ from one effort to another. The Job Plan represents a concerted effort to furnish the best answers to the following key questions:

What is it?
What does it do?
What must it do?
What does it cost?
What is it worth?
What else might do the job?
What does the alternative cost?
What will satisfy all of the agency's needs?
What is needed to put the VE change into effect?

Information Phase

This first phase of the Job Plan has three objectives.

- Understanding the product being studied.
- Determining essential functions.
- Estimating the potential value improvement.

Both the product itself as well as the general technical areas that it represents must be studied in order to understand the subject of the VE effort. Data accumulated should include the predicted total cost of ownership; the present configuration; the quality, reliability, and maintainability attributes; the quantity involved; and the background information regarding the reason that the product was originally selected (one of the most important considerations of all). It is most important to find qualified sources to obtain facts, not opinions. All data must be supported with adequate documentation.

When submitting VE proposals to the public agency or the architect-engineer, the contractor must make certain that the proposal is complete, including all documentation required. The proposal must include a functional analysis, an economic analysis, and a cost analysis to respond to the requirements for submittals of VE proposals.

Speculation Phase

The purpose of the speculation phase is to formulate alternative ways of accomplishing the essential functions. The effort begins as soon as enough information has been gathered, reviewed, and understood. Four of the methods used to help answer the basic question raised in this phase "What else might do the job?" are:

Simple comparison A thorough search for other items that have at least one significant characteristic similar to the item being studied.

Functional comparison A creative session in which new and unusual contributions of known items or processes are combined and rearranged to provide different ways of achieving the function.

Scientific factors A search for other scientific disciplines capable of performing the same basic function. This involves contributions from specialists in disciplines not normally considered in the original design.

Create and refine Push to get off the beaten path and reach for an unusual idea or another approach. Refine by strengthening and expanding ideas that suggest a different way to perform the function.

Analysis Phase

The purpose of the analysis phase is to select for further refinement the most promising of the alternatives that were developed during the speculation phase. During the speculation, there was a conscious effort to delay judgment so that the creative process would not be inhibited. During this phase, those ideas should be subjected to a preliminary screening to identify those that satisfy the following criteria:

Will the idea work?
Is it less costly than the present design?
Is it feasible to put into effect?
Will it satisfy the agency's needs?
If the answer to any of these is no, can it be modified or combined with another idea to give a yes answer?

The ideas that survive the initial screening are then rated according to their relative ability to meet the requirements just presented. The advantages and disadvantages of each are also noted. Preliminary cost estimates should then be developed for those ideas that appear technically and economically most promising. Following these preliminary estimates, one or more of the ideas with significant savings potential should be selected for further detailed analysis. If relative cost differences among several alternatives are not conclusive at this point, they may all be subject to further analysis.

Development Phase

In the development phase, the alternatives that have survived the previous selection process should be developed into firm recommendations (VE proposals). This portion of the effort includes the development of detailed technical and economic data for the proposal.

The VE proposal should contain everything, including total ownership costs. The development phase should also be devoted to assuring that the proposal satisfies all of the agency's needs. For an evaluation of mechanical equipment on a project, a checklist such as the following is often helpful:

Performance requirements.
Quality requirements.
Reliability requirements.
System compatibility.
Safety requirements.
Maintenance considerations.
Logistic support problems.

At this stage, the value engineering proposal should also include a written discussion designed to satisfy any objections likely to be raised concerning any aspect of the proposal. Conferences with specialists are often helpful in overcoming anticipated objections in advance. If a technical characteristic is either unacceptable or marginal, the alternative should be modified to correct the deficiency whenever possible. If it is not possible to overcome the deficiency, the alternative should be rejected. Of the technically feasible alternatives remaining, only the one that represents the lowest cost is selected for the detailed development of the technical and economic data necessary to support its selection.

If more than one alternative offers the possibility of valid savings, it is common for all of them to be recommended. One becomes the primary recommendation, the others are considered alternative recommendations, and all are usually presented in the order of their decreasing savings potential.

Presentation and Followup Phase

The final phase of the Job Plan includes the preparation and presentation of the value engineering proposal (VEP) to those having approval authority. This phase includes:

Preparing a plan for implementation.
Obtaining a decision regarding the disposition of a proposal.
Assisting as needed in the implementation actions.
Preparing a final report if appropriate.

When finally presented for approval, the VE proposal should be self-explanatory and leave no doubt concerning its justification. Only factual and relevant information should be included. All expected technical and cost variations from the existing design must be described. The following checklist represents the minimal information usually included in a value engineering proposal:

Identity of the project.
"Before" and "after" descriptions.
Cost of the original design.
Cost of the proposed design change.
Quantity basis for costs.
Cost to put VE proposal into effect.
Expected savings to agency.
Actions necessary to put change into effect.
Proposed schedule to put change into effect.

The management personnel who are responsible for review and approval must base their judgment entirely upon the documentation submitted with a VE proposal, and this documentation must contain all the data that a reviewer will need to reach a decision. VE proposals should contain sufficient information to assure the reviewer that:

   The performance of the overall project is not adversely affected.
   Supporting technical information is complete and accurate.
   Potential savings are based upon valid cost analyses.
   Feasibility of the proposed change is adequately demonstrated.

Failure to provide adequate VE proposal documentation is a major cause of proposal rejection. An analysis made of 90 rejected contractor-initiated Value Engineering Change Proposals (VECP) revealed that approximately 40 percent of the rejections were due to incomplete or inaccurate technical or cost information.

Construction Claims

CHAPTER · 11 ·

- Claims and Disputes
- Legal Counsel
- Differences Between the Parties
- Subcontractor Relations
- Documentation for Claims
- Resolution of Construction Disputes
- Supporting the Contractor's Position
- Alternative Methods for Dispute Settlement

CHAPTER 11 CONSTRUCTION CLAIMS

Not all public agencies will recognize or be willing to give the contractor the benefit of the doubt. More probably, they will generally assume that the contractor is only one step away from being an unprincipled charlatan, whose only motive is to bid low to get the job and then inundate the agency with claims for extras, extensions of time, differing site condition claims, and numerous other innovative methods of separating them from more money than the agency planned to pay.

Then also there are those public agency officials who appear to be on a crusade to drive the contractor into bankruptcy, as if they needed any help to accomplish this.

Needless to say, the time will come, probably early in a job, when you reach your first impasse and know that you are being unfairly treated in a matter involving time or money on the job. After repeated attempts to negotiate in good faith, there appears to be only one avenue open to you: the submittal of a claim.

In this chapter, the authors will outline some recommendations about the preferred course of action before making a demand for arbitration or litigation, and, failing that, some suggestions on methods of recovery of your costs on your claim.

CLAIMS AND DISPUTES

Whether it is the competitive bidding process, increased competition, or just part of the growing trend toward more litigation, more and more projects are being affected by claims and disputes. Very often, a majority of such claims and disputes arise out of poorly drafted or ambiguous contract documents. Disagreements between the contractor and the agency, and increasingly including the architect-engineer, are becoming more common; these disagreements usually regard the interpretation of the plans and specifications, what should be considered extra work in a contract, payment for contract work and change orders, extensions of time, damages for delay caused by either the agency or the contractor, changed or unforeseen conditions, performance of subcontractors, compliance with contractual requirements, partial acceptance of a project by the agency, and other similar problems. When such claims or disputes arise, a contractor must pay careful attention to the procedures set forth in the contract documents to handle claims and disputes.

Contractor Must Alert Agency

Typically, contract documents require a contractor to alert the agency and/or the architect-engineer immediately, or within a certain number of days, to any potential claim or dispute when it arises. Contracts normally require that any claim or dispute be submitted to the architect-engineer or agency's representative as the first step in the claims resolution process. Contracts often set forth an appeals procedure so that the contractor can proceed with its claim or dispute if the architect-engineer's decision is unfavorable. When all the administrative steps set forth in the contract documents have been exhausted, the contractor may then, and only then, resort to the courts or arbitration, if provided for under the contract. Whatever procedural steps are set forth in the contract must be followed by the contractor. Failure to follow these steps could in itself be considered a breach of the contract by the contractor.

Administrative Procedures

Any claim or dispute that the contractor has will necessarily be filed against a governmental agency. As such, a contractor must exhaust all of its administrative remedies before resorting to the courts and must strictly adhere to

CHAPTER 11 CONSTRUCTION CLAIMS

any claim procedures mandated by law. For example, in addition to the claims and disputes provisions outlined in the typical public agency contract, some states require that a government code claim be filed by the contractor against the appropriate agency, followed by that agency subsequently rejecting the claim within a prescribed time, the contractor then has the right to take the matter to court. A contractor's failure to follow mandated government claims procedures can result in the contractor being barred from recovery on its claim.

Work Performed Under Protest

Except for total nonpayment by the agency or particular circumstances covered under a specific contract, the contractor cannot refuse to proceed with the work on a project when a claim or dispute arises. Normally, the contractor must continue its operations despite any claims or disputes. Generally, the contractor must perform any disputed work under protest and still complete the project in question on schedule, leaving any unresolved claims or disputes until after the project has been completed.

When a claim or dispute arises or when the contractor foresees that it will arise, the contractor must take two immediate steps. First, the contractor must give the agency notice of any claim or dispute, or any potential claim or dispute, as soon as possible. Just as it is often said that the most important considerations when planning to invest in real estate are "location, location, and location," the three most important considerations when making a construction claim are "notice, notice, and notice." Most contracts contain very specific requirements for Notice of Claims or Notice of Potential Claims that set forth specific time frames within which a contractor must notify the agency. Although the courts have varied in their interpretation and enforcement of such time requirements, a contractor runs an unnecessary risk when claims are not made in accordance with contractual requirements.

Agency Must Have Opportunity to Correct

Regardless of any contractual language regarding the time for making claims, common sense and equitable principles dictate that once a contractor becomes aware of a claim or a potential claim, it must be communicated to the agency. For example, if the agency's actions in some way delay the contractor and the project, the agency must be advised of the situation in order for the agency to have an opportunity to correct or alleviate the problem. If a contractor does not give notice of a delay to the agency at the time it is actually occurring, the contractor will have very little chance of recovering damages incurred as a result of such a delay after the project is complete. Without proper notice, the agency will argue, and rightfully so, that it cannot be held liable to the contractor for a delay that it was never aware of or had an opportunity to do anything about.

Preparation of a Claim File

The second step that a contractor should take is to begin a claim file, documenting in full all aspects of its claim. Correspondence, photographs, job logs, separate and detailed cost records including cost isolation data, estimates, schedule networks, and any other pertinent documentation should be made a part of the file. The better and more complete such claim records are, the greater the chance of a favorable resolution to a claim, either by way of settlement or litigation.

The Right to File Claims

Contractors are entitled to and should feel free to make any real and provable claim that arises during the course of a project. Claims should not be made or taken "personally," as they are simply the means available to the parties to the contract to be able to adjust the contractual and economic relationship between them to meet changing conditions. However, claims should not be viewed by the contractor as a way to make up for a bad bid, or to recoup losses incurred on the project. Ultimately, it is the integrity of the contractor that dictates what claims are and are not made against an agency. In general, if a contractor only presents claims to an agency that are legitimate and verifiable, it will stand a better chance of having those claims, and all other claims it files, settled in a manner favorable to the contractor. If, on the other hand, the contractor makes it a practice to file claims for any and all reasons, some less legitimate than others, the contractor runs the risk of prejudicing its legitimate claims and reducing the possibility of resolving them without litigation, a costly and delayed process.

A contractor should pursue every legitimate claim that it has, but recognize that the agency is entitled to proper notice and adequate proof of the claim before it has any obligation to pay on the claim. Contrary to a bidding situation in which the contractor assumes only the best case circumstances, a claim should be estimated assuming only the worst case circumstances when a contractor finds it necessary to include impact costs yet to be incurred.

Particular attention should be paid to situations in which claims are directly related to the actions or inactions of the agency's representatives or agents. Claims that arise as a result of the obligations of the agency's architect-engineer to properly design a portion of a project are assertable against the agency, and under some circumstances, directly against the architect-engineer. In light of the very close relationship that normally exists between the architect-engineer and the agency on a project, the contractor must be especially diligent and thorough when making a claim that the contractor believes has been caused by the architect-engineer.

Although claims often provide the contractor with an opportunity to make up for mistakes made in its bid or the construction of the project, it is not good practice to make more claims simply to compensate for mistakes and errors. Mistakes are made, and errors do occur. A contractor must accept this fact and continue. Construction will go on, and there will be more jobs and opportunities to compensate for such mistakes and errors in the future. No purpose will be served by turning a project into a battleground in order to cover losses incurred as a result of a contractor's own mistakes and errors.

LEGAL COUNSEL

Normally, a contractor does not involve an attorney in a particular project until a claim or dispute is in the litigation stage, which is often well after the project has been completed. At that point, the attorney only has the facts, documents, and records that exist to work with, and the attorney must essentially play the game with the cards that have been dealt. Had the attorney been brought in earlier when the claim first arose, it may have been possible to help the contractor's case from the beginning and put the contractor in the strongest possible position. Had the attorney been involved with the project from the very beginning, even prior to signing the contract, the claim itself might have been avoided.

Construction cases, whether litigated in court or arbitrated, usually involve hundreds and hundreds of facts, numerous documents of many different types, multiple parties, and many witnesses. Most cases involve conflicting issues of liability and damages. As a result, construction litigation is considered the second most complex type of litigation, after antitrust litigation. It is very expensive in terms of both time and money.

Involvement from the Project Beginning

For these reasons, it is wise for a contractor to include a lawyer familiar with and experienced in construction law in the jurisdictional area where the project is located as a member of the project "team," along with its bonding agent and insurance broker. By involving legal counsel in a project from the beginning, the contractor will have a better chance of making it a profitable job. A lawyer's review of the prime contract documents and the subcontract documents will help identify the legal ambiguities contained in them and the specific obligations that the contractor has under their terms. By being able to anticipate problems, the contractor will be better able to avoid or at least plan for them.

When a claim situation arises, the contractor will have legal counsel available to advise it of its potential liability and to help structure and prepare a claim. A lawyer's intimate involvement with a project from its outset is an invaluable asset when a claim ultimately goes to court or arbitration.

DIFFERENCES BETWEEN THE PARTIES

In general, most claims issues will fall into approximately 10 classifications, with numerous variations possible within each category.

1. Owner-caused delays in the work.
2. Owner-ordered scheduling changes.
3. Constructive changes.
4. Differing site conditions.
5. Unusually severe weather conditions.
6. Acceleration of the work; loss of productivity.
7. Suspension of the work; termination.
8. Failure to agree on change order pricing.
9. Conflicts in plans and specifications.
10. Miscellaneous problems.

Owner-Caused Delays

The majority of all claims involve at least some elements of delay, even if the primary issue is one of the other categories. Most contractors, until they have consulted a claims specialist, fail to realize the potential for recovery of losses in this sensitive claims area.

Delay claims fall into three categories: nonexcusable, excusable, or compensable. A nonexcusable delay is one that is caused by factors within the contractor's reasonable control. Essentially, that means the delay is the contractor's fault, so the contractor will be unable to recover additional time or compensation.

A delay that is caused by factors beyond the contractor's reasonable control, but not the result of the owner's actions or failure to act, is considered excusable. An excusable delay entitles the contractor to an extension of time, but no additional compensation for the cost of the delay (see Figure 11.1).

A delay that is considered compensable is one in which the owner has failed to meet an obligation stated or implied in the construction contract. If a delay is considered compensable, the owner must grant a time extension and reimburse the contractor for the increased cost caused by that delay. If a contractor experiences concurrent delays and one is compensable and the other merely excusable, no compensation is allowed. Similarly, on concurrent delays when one is caused by the contractor and the other by the owner, the delay is considered neither excusable nor compensable to the extent of

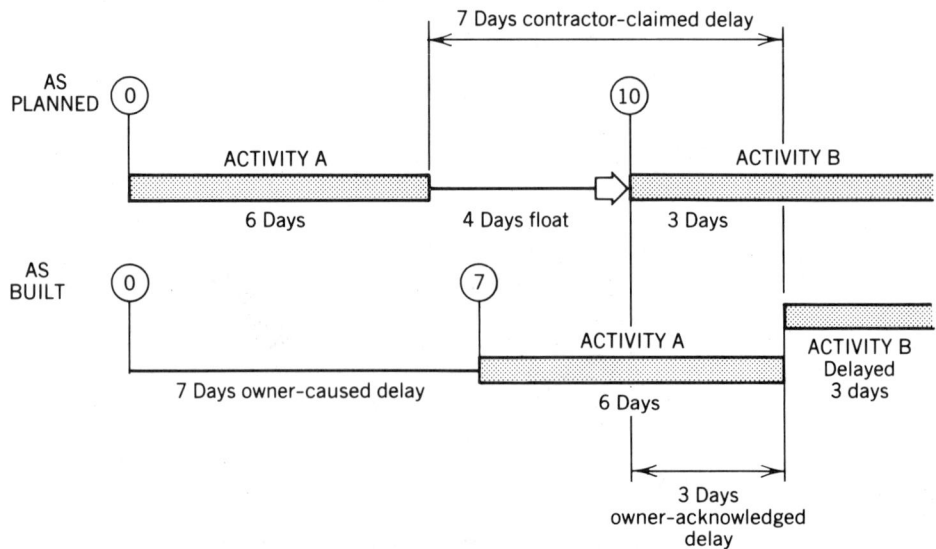

FIGURE 11.1 Example of disputed time in owner-caused delay.

the overlap or concurrency. [Appeal of Rivera General Contracting ASBCA No. 25888 (April 30, 1985)] It should be stressed that these definitions are general in nature. The parties involved have great latitude to contractually determine whether or not the contractor will be entitled to additional compensation, an extension of time, or nothing at all.

Delays are the most prevalent problem on construction projects. A variety of factors in an on-going project can lead to a delay. Typical of the types of delays that might be caused by the public agency include the following:

1. Late approval of shop drawings and samples.
2. Late approval of laboratory tests.
3. Delays in answers to field inquiries by the contractor.
4. Changes to the contractor's method of doing the work.
5. Variation in estimated quantities.
6. Interference with the contractor during construction.
7. Owner-caused schedule changes.
8. Design changes.
9. Changes in inspection level.
10. Failure to provide for site access.
11. Lack of required rights-of-way.
12. Interference by other contractors or owner's forces.

A contractor can jeopardize its claim, however, by being the cause of nonexcusable delays of its own. Such delays are frequently the result of any of the following causes:

1. Late submittal of shop drawings.
2. Late procurement of materials or equipment.
3. Insufficient personnel.
4. Unqualified personnel.
5. Inadequate coordination of subs or other contractors.
6. Subcontractor delays.
7. Late response to agency/architect-engineer inquiries.
8. Construction not conforming to contract requirements making repeated reworking necessary.

Unabsorbed Home Office Overhead

The costs of delays involve many elements. Naturally, the direct costs are affected, but a contractor's work efficiency, the construction schedule, available favorable weather, impact costs, and even home office overhead may also be affected. Normally, most contract documents specify the limits of overhead charges that may be levied for force account work, but do not give up just yet; it should be kept in mind that such limitations apply only to a contract that is completed within the specified time. If the public agency causes a compensable delay that forces the contract into a longer time frame, the contractor is entitled to unabsorbed or extended home office overhead.

The most common way of computing the value of extended home office overhead is based on the use of the "Eichleay formula." The validity of the Eichleay formula was successfully challenged in February 1983, but was reborn again a year later when the U.S. Court of Appeals for the Federal Circuit reinstated it. [*Capital Electric Co. v. United States,* Appeal No. 83-965 (February 7, 1984)] The Eichleay formula is a method of calculating home office overhead damages in delays, suspensions, or extensions of work. Such damages have usually been called "extended" home office overhead.

The Eichleay formula is commonly expressed as follows:

a. $\dfrac{\text{contract billings}}{\text{total billings}} \times \dfrac{\text{total overhead for contract period}}{} = \dfrac{\text{allocable overhead to the contract}}{}$

b. $\dfrac{\text{allocable overhead}}{\text{days of performance contract period}} = \text{daily contract overhead}$

c. $\text{daily contract overhead} \times \text{number of days delay} = \dfrac{\text{amount claimed (total added overhead)}}{}$

"Unabsorbed" home office overhead is computed in a similar manner, but relates only when the direct cost base is not large enough to absorb the fixed overhead (an indirect cost) at the contractor's "normal" absorption rate. Fixed overhead is an indirect cost in that it is a cost incurred for the benefit of more than one cost objective or project. Direct costs, of course, are always identified with separate projects. However, indirect costs cannot be identified in this manner, and as a result, they must be allocated or distributed in a logical manner among all of the contractor's projects.

A definition may be helpful at this point. Unabsorbed overhead exists when the direct cost base is not large enough to absorb the fixed overhead (an indirect cost) at the contractor's normal absorption rate. Fixed overhead is an indirect cost that is incurred for the benefit of more than one project. By nature, direct costs are identified with separate projects. However, indirect costs are not normally identified with separate projects and must therefore be allocated, or distributed, in a logical manner among the several projects that the contractor has going.

The allocation of indirect costs is usually done by using a certain "base," for example, labor cost, contract billings, machine hours, or a similar base. The Eichleay formula uses *contract billings* as a base. A ratio of the base for a specific project to the total base for all projects is used to allocate the indirect cost. This is called "absorption costing"—the distributed cost is absorbed by the projects.

One of the problems with the Eichleay formula as a means of calculating underabsorbed home office overhead is that it does not reflect underabsorption of overhead. If we assume that a contractor can offer proof of entitlement, the Eichleay formula results in recovery regardless of whether there was an actual underabsorption of home office overhead. For example, if the delay is caused by the influx of a large amount of extra direct costs as a result of numerous change orders, the contractor's base for absorbing home office overhead may actually increase to the extent that overhead is overabsorbed despite the extended duration of the project. In other words, the contractor was not prevented from obtaining additional work and forced to allocate an unfair amount of overhead to the project. Instead, the contractor may have obtained plenty of additional work through additional change orders to which the contractor may have been able to allocate home office overhead.

As an example, consider a fictitious project we shall call Project A. If work on Project A is suspended, the contractor cannot earn any revenue, and its billings decrease. However, during the suspension period, the contractor still continues to incur the fixed overhead that was originally allocated to Project A prior to its suspension. This fixed overhead for the now suspended Project A is "unabsorbed" during the suspension period; there is no Project A revenue to absorb these overhead costs. In other words, the unabsorbed overhead of Project A must now be absorbed by the remaining projects. However, since the actual allocation base during suspension is less than that of the as planned base, the remaining projects must support more overhead than was originally planned, or is "normal." For example, say that the normal overhead rate is 10 percent for all projects. Without Project A, it increases to 15 percent for the remaining projects. The suspension causes the overhead rate to increase. In this example, the overhead rate "differential" is a 5 percent increase.

It should be pointed out that "overabsorption" can also occur, such as when the base increases. For example, say that the suspended work of Project A is performed during a later period. Fixed overhead remains as planned, but the billings *increase*. Thus, the allocation base is higher than the as planned base. Then, although the overhead rate may have been 10 percent, with a larger base it may actually drop to 5 percent. The overhead rate differential is then a 5 percent *decrease*. This is called *overabsorption* of fixed overhead.

Because the damages computed through the Eichleay method bear no relationship to actual absorption rates, contractors with home office overhead claims may be forced to rely upon other methods, such as the two examples of construction-delayed projects illustrated in Figures 11.2 and 11.3. The first method is known as the *comparative absorption rate* method. Under this method, you determine the underabsorbed overhead by finding the difference between overhead actually incurred and overhead that would have been incurred if the contractor had been able to maintain a reasonable absorption rate. The second method is known as the *burden fluxuation* method. It has been used by Boards of Contract Appeals in calculating manufacturers' underabsorbed overhead claims. This method determines underabsorbed overhead by finding the increase in absorption rate and allocating that increase to work on other projects that were forced to bear more than their fair share of overhead expenses. An Eichleay calculation for each of the two examples is provided for further comparison.

Example No. 1 (Figure 11.2) illustrates a project that could have been performed by the contractor for $400,000 over a four-month duration, if we assume no change orders were issued and no suspensions of work or other delays were encountered. The contractor in this example has a fixed home office overhead rate of $40,000 and regularly does $500,000 worth of total business per month, including the subject project. These data are reflected under the heading "potential performance" and indicate what would have happened if there had been no changes or delays caused by the owner on this project.

In the actual performance of the work, the contractor experienced no changes; however, there was a one-month suspension of work during the third month of the project. This resulted in a one-month delay to the contractor in question and the inability to take on $100,000 in new work after the planned completion date had passed. Under the Eichleay theory, the

Example No. 1: A $400,000, four-month contract with a one-month suspension and no change orders.

	MO. 1	MO. 2	MO. 3	MO. 4	MO. 5
POTENTIAL PERFORMANCE					
Home office overhead	$ 40,000	$ 40,000	$ 40,000	$ 40,000	$ 40,000
Contract billings	$100,000	$100,000	$100,000	$100,000	—
Other contract billings	$400,000	$400,000	$400,000	$400,000	$500,000
Total billings	$500,000	$500,000	$500,000	$500,000	$500,000
ACTUAL PERFORMANCE					
Home office overhead	$ 40,000	$ 40,000	$ 40,000	$ 40,000	$ 40,000
Contract billings	$100,000	$100,000	—	$100,000	$100,000
Other contract billings	$400,000	$400,000	$400,000	$400,000	$400,000
Total billings	$500,000	$500,000	$400,000	$500,000	$500,000

METHOD NO. 1: EICHLEAY FORMULA

$$\frac{\$400,000 \text{ (Contract billings)} \times \$200,000 \text{ (Total overhead)}}{\$2,400,000 \text{ (Total billings)}}$$

$$= \$33,333 \text{ (Allocable overhead)}$$

$$\frac{\$33,333 \text{ (Allocable overhead)}}{5 \text{ (Months of performance)}} \times \$6,667 \text{ (Monthly contract overhead)}$$

$$\times \text{ 1-Month delay} = \$6,667 \text{ claim}$$

METHOD NO. 2: COMPARATIVE ABSORPTION RATES

$$\frac{\$200,000 \text{ (Potential total overhead)}}{\$2,500,000 \text{ (Potential total billings)}} = 8\% \text{ (Reasonable overhead ratio)}$$

$$8\% \text{ (Reasonable overhead ratio)} \times \$2,400,000 \text{ (Actual total billings)}$$

$$= \$192,000 \text{ (Reasonable total overhead)}$$

$$\$200,000 \text{ (Actual total overhead)} - \$192,000 \text{ (Reasonable total overhead)}$$

$$= \$8,000 \text{ Underabsorbed overhead claim}$$

METHOD NO. 3: BURDEN FLUXATION

$$\$2,400,000 \text{ (Total billings)} - \$400,000 \text{ (Contract billings)}$$

$$= \$2,000,000 \text{ (Other billings)}$$

$$\frac{\$200,000 \text{ (Actual overhead rate)}}{\$2,400,000} - \frac{\$200,000 \text{ (Potential overhead rate)}}{\$2,500,000} =$$

8.33% — 8.00% — 0.33% (Burden fluctuation)
0.33% × $2,000,000 = $6,600 Underabsorbed overhead claim

FIGURE 11.2 Comparative absorption rate method of computation. (*Source*: Reprinted with permission by Phillip R. McDonald, Esq., for *Construction Claims Monthly*, 951 Pershing Drive, Silver Spring, MD 20910.)

contractor would have a claim for only $6667 of extended overhead, based upon the one-month delay.

A comparison of absorption rates reveals that the contractor, in fact, suffered $8000 in underabsorbed overhead during the delay. This figure was arrived at by calculating a reasonable overhead rate, based upon what the contractor's overhead absorption rate would have been, had it not been for the delay. In this case, the rate equals 8 percent. By applying that rate to the actual billings for the period involved, the contractor can arrive at a reasonable total overhead of $192,000, which is $8000 less than the amount of overhead that the contractor had to absorb. In other words, the contractor was unable to maintain the 8 percent absorption rate during the delay. Had the contractor been able to maintain that rate, only $32,000 would have been expended on home office overhead during the delay.

Under the *burden fluxuation* method, the contractor would claim a 0.33 percent increase in its overhead rate. By applying that percentage increase to the work that had to bear the extra overhead cost, the contractor could claim $6600. Regardless of the method used, this example presents a classic case of the underabsorption of home office overhead. Recovery, however, would be clearly dependent upon proof that additional work could have been obtained in other contracts, had it not been for the delay on this project.

Example No. 2 (Figure 11.3) is based upon the same contract and contractor described in Example No. 1, but on a different type of delay. The actual performance included no suspension of work and resulted in the increase of the contract price by $110,000 for change orders, including $10,000 of overhead markup. Under the Eichleay formula, the contractor would have an extended home office overhead claim of $8127 despite the fact that change orders resulted in additional direct charges that could absorb the contractor's overhead.

Calculating the contractor's damages by comparing absorption rates shows that the contractor actually did not suffer from underabsorption of home office overhead. Instead of absorbing the overhead at the normal rate of 8 percent, the contract, with the help of changes, was able to absorb overhead at the rate of 7.97 percent. This resulted in $800 of overabsorbed overhead. Moreover, if the $10,000 markup on the changes was required to be set off against the contractor's claim, the contractor would have actually absorbed its overhead at the rate of 6.8 percent. The burden fluxuation method would similarly reveal an overabsorption of overhead in the amount of $600, based upon the 0.03 percent decrease in the contractor's overhead rate.

Example No. 2: A $400,000, four-month contract with no suspension of work and change orders worth $110,000 including $10,000 overhead markup. Potential performance is the same as in Example No. 1.

	MO. 1	MO. 2	MO. 3	MO. 4	MO. 5
ACTUAL PERFORMANCE					
Home office overhead	$ 40,000	$ 40,000	$ 40,000	$ 40,000	$ 40,000
Contract billings	$ 75,000	$ 75,000	$125,000	$135,000	$100,000
Other contract billings	$400,000	$400,000	$400,000	$400,000	$400,000
Total billings	$475,000	$475,000	$525,000	$535,000	$500,000

METHOD NO. 1: EICHLEAY FORMULA

$\dfrac{\$510{,}000 \text{ (Contract billings)}}{\$2{,}510{,}000 \text{ (Total billings)}} \times \$200{,}000 \text{ (Total overhead)}$

$= \$40{,}637 \text{ (Allocable overhead)}$

$\dfrac{\$40{,}637 \text{ (Allocable overhead)}}{5 \text{ (Months of performance)}} \times \$8{,}127 \text{ (Monthly contract overhead)}$

$\times$ 1-Month delay = $8,127 claim

METHOD NO. 2: COMPARATIVE ABSORPTION RATES

$\dfrac{\$200{,}000 \text{ (Potential total overhead)}}{\$2{,}500{,}000 \text{ (Potential total billings)}} = 8\% \text{ (Reasonable overhead ratio)}$

8% (Reasonable overhead ratio) $\times$ $2,510,000 (Actual total billings)

= $200,800 (Reasonable total overhead)

$200,000 (Actual total overhead) − $200,800 (Reasonable total overhead)

= ($800) Underabsorbed or $800 overabsorbed overhead

METHOD NO. 3: BURDEN FLUXATION

$2,510,000 (Total billings) − $510,000 (Contract billings)

= $2,000,000 (Other billings)

$\dfrac{\$200{,}000 \text{ (Actual overhead rate)}}{\$2{,}510{,}000}$ − $\dfrac{\$200{,}000 \text{ (Potential overhead rate)}}{\$2{,}500{,}000}$ =

7.97% − 8.00% = (0.03%) (Burden fluctuation)

(0.03%) $\times$ $2,000,000 = ($600) Underabsorbed or $600 overabsorbed overhead

FIGURE 11.3 Burden fluxuation method of computation. (*Source:* Reprinted with permission by Phillip R. McDonald, Esq., for *Construction Claims Monthly*, 951 Pershing Drive, Silver Spring, MD 20910.)

The results obtained for the two examples through the absorption rate or burden fluxuation methods vary greatly, but the results obtained by using the Eichleay formula do not. This illustrates some of the objections to the use of the Eichleay method. The contractor in Example No. 1 that experiences a one-month suspension without receiving any extra work clearly suffers more damages than the contractor in Example No. 2 that receives extra work extending its contract through changes; yet, the Eichleay formula would give the second contractor a larger recovery.

Flaws in Use of the Various Computation Methods

Several criticisms can be raised about the burden fluxuations method. Some of these criticisms also apply to the basic Eichleay approach. First, the use of a differential is not a cost accounting method, but an approximation method that is based upon certain assumptions.

Burden fluxuation assumes that all of the overhead rate differential is due to the one impacted project. It does not consider other factors that may impact a contractor's overhead costs, such as other souring jobs that a contractor may have. It may therefore overcompensate the contractor.

The normal rate is an *averaged* rate for all jobs. Different projects may, in fact, have overhead rates lower or higher than the average rate. How, then, is the normal rate for a particular job calculated? Even knowing the contractor's cost records and the nature of the indirect costs, it is difficult to find a way to know *directly* which part of a fixed overhead differential is due to impacts, delays, suspensions, extensions, and the like of one particular project.

Second, the unabsorbed overhead determined by the burden fluxuation method would be expected to be reduced by the amount of money received by the contractor on change orders.

Third, the burden fluxuation method does not account for the growth-decay bell curve of costs. This also applies to the Eichleay formula.

Fourth, the method does not account for variable overhead costs. Burden fluxuation treats all overhead costs as fixed, thereby introducing some distortion.

Fifth, some home office overhead costs may be directly attributable to specific jobs or could be distributed on some percentage-of-effort basis instead of by an allocation formula.

Sixth, a more appropriate base than total costs may be found to allocate overhead. One suggestion might be to use labor costs as a base. This would have the effect of overcompensation for a contractor on a labor-intensive job, or undercompensation for a contractor on a material-intensive job.

Thus, there are problems with the burden fluxuation method, and the Eichleay formula has some of the same problems. However, burden fluxuation does answer in part those complaints that the daily rate method of Eichleay is not related to actual costs. Through the use of actual overhead rates, the burden fluxuation method is more closely tied to actual costs.

Scheduling Changes

Any scheduling change can have an important effect on the contractor's operations. When scheduling changes are implemented as a result of constraints created by the owner or architect-engineer, the basis for a contractor claim may exist.

Whenever the owner or the architect-engineer issues any change order, field order, or directive, or whenever a constructive change condition exists in any of the following areas, the contractor should begin to consider the impact of such action by the agency or architect-engineer upon project costs and profitability.

The areas of primary sensitivity include agency- or architect-engineer-caused delays, requirements to deviate from the schedule, orders to expedite the work, job interference by the agency or architect-engineer, agency constraints on scheduling interfaces, late availability of agency-furnished material, agency-imposed acceleration or deceleration, impractical or impossible milestones, extra work, schedule impacts, sequence of construction by the owner or architect-engineer, changes in completion dates, time extensions, utilization of scheduling float time, and schedule approvals. If a CPM schedule is not available during the job, any consistent scheduling method that illustrates the delay will help support the claim.

Any change in the schedule imposed by either the agency or the architect-engineer can become the basis for a potential claim by the contractor. The best way to assure a sound, defensible claim is to utilize CPM-scheduling techniques. This form of scheduling documentation offers the greatest protection.

As a job progresses, time extensions may be granted by the owner or architect-engineer. These would result in changes to the schedule and could result in revised interface dates, or changed completion dates. The contractor should keep a close watch on the status of the schedules that it submits to the agency or architect-engineer for approval. Departures in the contractor's schedule from the schedule originally intended by the owner may be considered as contract amendments after approval by the agency. If a contractor submits a shorter schedule than that originally called for under the contract, and it is approved by the agency or its architect-engineer, any liquidated damages provided under the contract may now be applied to the early finish date shown in your own shortened schedule.

There is no question that the only way to make a profit on a project is to get in and get out in the shortest possible time, thus reducing the home office overhead allocated to that project. But, that is no reason to submit a short schedule to the owner. If you do so, you are simply making your own position less flexible. A long allowable schedule by the agency is good insurance against an unforeseen delay by your suppliers or subcontractors. One major contractor known to the author submitted a two-year construction CPM to the owner with three months of float time at the very end of the project to avoid forfeiting the time. Ridiculous? Certainly, in principle, but as a matter of fact, unforeseen noncompensable delays forced this contractor into a position of needing every day of that terminal "float."

Constructive Changes

A constructive change is an informal act authorizing or directing a modification to the contract caused by the agency or architect-engineer through an act or failure to act. In contrast to the mutually recognized need for a change, certain acts or a failure to act by the agency that increases the contractor's cost and/or time of performance may be considered grounds for a change order. This is termed a *constructive change,* however, it must be claimed in writing by the contractor within the time specified in the contract documents. Otherwise, the contractor may waive its rights to collect. Types of typical constructive changes may include

- Defective plans and specifications.
- Architect-engineer's interpretations of documents.
- Higher standard of performance than specified.
- Improper inspection and rejection.
- Change in the method of performance by owner.
- Change in construction sequence by owner.
- Owner nondisclosure of pertinent facts.
- Impossibility or impracticability of performance.

An example of a constructive change resulting from failure to act is the Appeal of Continental Heller Corp. [GSBCA No. 7140 (March 23, 1984)]. In this case, the government's failure to grant a legitimate request for a time extension was held to be a constructive acceleration of the work schedule.

Continental Heller Corp. was awarded a contract for the construction of a federal building in San Jose, California. Heavy rains at the start of the project made it impossible for the excavation subcontractor to proceed.

Nevertheless, the contracting officer insisted that the contractor stay on schedule and refused to grant a time extension until the contractor documented both the site conditions and the status of the activity on the critical path. An extension was finally granted 16 months after completion of the excavation. In the meantime, however, the subcontractor had switched to a more expensive method of excavation in order to remove the saturated soil in accordance with the original schedule.

The Board of Contract Appeals of the General Services Administration found classic elements of a construction acceleration case. The delay was excusable, yet the government forced the contractor to adhere to the original performance period, thereby causing the contractor to incur additional costs. Continental Heller was awarded $113,165 plus interest.

The Board of Contract Appeals, referring to the belated time extension granted to the contractor 16 months later, said

> As a defense to a claim of constructive acceleration, a belated time extension is worthless . . . . It had to have been clear to anyone who did not sleep through the entire two days that the soil at the site was saturated with moisture and could not be compacted as required. The government could not, by continually insisting on documentation of what was already known, justify its refusal to grant a time extension.

Differing Site Conditions

Sometimes referred to as "changed conditions" or "unforeseen conditions," the term *differing site conditions* is typically used in all federal contracts, and there is a growing trend by many public agencies to adopt similar wording.

Read the contract documents carefully. Somebody prepared them for a reason and it probably was not to help the contractor. The failure of a public agency to provide payment for differing site conditions places the contractor in a difficult position. If the agency takes a hard line position on this issue, the contractor may find it necessary to seek relief from a court, a process that is both lengthy and costly.

The federal policy is to make adjustments in time and/or price when unknown subsurface or latent conditions at the site are encountered by the contractor. The purpose is to have the agency accept certain risks and thus reduce the large contingency amounts in bids to cover such unknown conditions. The federal government and many local agencies include provisions in their construction contracts that will grant a price increase and/or time extension to a contractor that has encountered subsurface or latent conditions.

Under the definitions for Differing Site Conditions offered in Chapter 9, an existing underground pipeline that either was not shown on the drawings at all, or that was incorrectly located on the contract drawings, would be a Type-1 Differing Site Condition. Unusually severe weather conditions for the time of year and the location of a given project may well fall into a Type-2 Differing Site Condition. The discovery of expansive clays in the excavation area, if not normally encountered in the location where the project was planned, and if not detected during soils investigations, may be either a Type-1 or -2 condition, depending upon circumstances. However, the discovery by the contractor of a permafrost condition in the tropics, for example, would most certainly be a Type-2 Differing Site Condition.

Unusually Severe Weather Conditions

There should be a word of caution about weather. Severe rains or similar weather conditions that prevent work from being done, or which in any way delay the project, may not always be excusable delays (see Figure 11.4), and in some cases, may be ruled excusable only and not compensable.

In the appeal of Inland Construction, Inc., [ENG BCA No. 5033 (October 11, 1984)] a contractor was denied a differing site condition claim because the Board found that the increased costs were caused by an unusually severe rainfall, an event "which only entitles the contractor to an extension of time." In another case, although the contractor was delayed by rain, its claim of excusable delay was denied because the severity of the storm could have been foreseen by the contractor [Appeal of B.D. Click Co., Inc., ASBCA No. 24586 (July 27, 1984)]. In that case, the Board of Contract Appeals ruled that weather delays are excusable only when the weather is abnormal. In the words of the Board,

> No matter how severe or destructive, if the weather is not unusual for the particular time and place, or if the contractor should have reasonably anticipated it, the contractor is not entitled to relief.

FIGURE 11.4 A severe rain may not always qualify as an excusable delay.

Nevertheless, the contractor should document *all* weather delays as they occur. The form in Figure 11.5 provides a good tool for this. The determination of compensability can be made at a later date.

Acceleration of the Work by the Public Agency/Owner

In considering acceleration, the first consideration must be whether the so-called "acceleration" is an order from the agency to a contractor that is behind schedule to get back on schedule, or an order to a contractor either directly or constructively to complete the work prior to the scheduled completion date.

A related condition called "deceleration" can also be experienced on a project. This occurs when a contractor is directed in writing or constructively to slow down its job progress. Many of the same considerations that apply to acceleration also apply to deceleration.

In preparing a claim for acceleration or deceleration, it should be borne in mind that the costs to the contractor for going into premium time, such as working an extended work week, cannot be computed simply as including the added hourly costs multiplied by the additional hours. Studies have shown that as the work week is extended, there is an accompanying loss in

B&H CONSTRUCTION **MEMORANDUM OF DELAY**

PROJECT __Arroyo Wastewater Treatment Plant__
JOB NO. __07-01-28675__

PROJECT MANAGER __Ray E. Barnes__

The undersigned have determined this date that a delay has occurred in proceeding with the controlling operation due to weather as defined and limited by the Contract General Conditions, on the following dates

Oct 15, 16, and Oct 19 through 23, 1987

for a total of __7__ working days.

The above dates are not included in any request of record, by the contractor, for a time extension, nor are they included in any time adjustment related to changed work. The contractor certifies that the controlling operation or operations adversely affected are:

Foundation and trench excavation

Trench backfill operations

Data used are fictitious for illustration only

Contractor __Kenneth MacDonald, Proj. Mgr.__

Resident Project Representative __Charles Dunham__

Distribution: 1. Proj. Mgr.
2. Field Ofc.
3. File

Wiley-Fisk Form 7-3

FIGURE 11.5 Documentation of weather delays at the site.

worker productivity. Furthermore, as the extended overtime is continued, the productivity rate continues to drop.

Suspension of the Work; Termination

The work on the project can usually be suspended by the agency for any one or more of several reasons. In each case, the contractor should keep detailed cost isolation records of all activities affected by the suspension. It should be kept in mind that suspension of the work for any amount of time such that the completion date is extended will impact the contractor's costs through unabsorbed home office overhead and the real possibility of missing other projects due to the delay.

The contractor should consider the effect upon its organization of the costs related to dismantling operations, mobilization or demobilization, direct costs, settlement expenses, escalation costs, prior commitments, post-termination continuing costs, unabsorbed overhead, unexpired leases, severance pay, implied agreements, restoration work, utility cutoff, inventory, replacement costs, and all other allocable costs. On suspensions of the work, be certain that all such orders are in writing, and that a careful record is kept of the order's total effect on the contractor's time and costs.

Failure to Agree on Change-Order Pricing

One of the most common causes of contractor claims occurs during attempts to price change orders. All too often, public contract change orders contain a waiver clause that requires the contractor to guarantee that the price and time named in each individual change order represents the total cost to the owner for that change, and the contractor waives any rights to impact costs.

Unfortunately, a contractor's only recourse is through the claims process. There is an encouraging note to the contractor, however, if we judge from the settlements ordered in at least one contract dispute. In the appeal of Centex Construction Co., Inc., [ASBCA No. 26830 (April 29, 1983)] the Armed Services Board of Contract Appeals ruled that the government may not force a contractor to "forward price" its impact costs when pricing a government-directed change.

In that case, Centex Construction Co., Inc. was a prime contractor on a military construction project. The government ordered a number of changes and asked Centex to submit cost proposals under the terms of its contract. Centex listed the *direct* costs of performing the changed work but did not list delay damages and other impact costs. When a change order was issued, Centex refused to release its right to submit a claim for impact costs.

Centex stated that it was impossible to accurately forecast such costs, particularly when an on-going series of changes is involved. The Appeals Board agreed with Centex, stating

> While it might be good contract administration on the part of the government to attempt to resolve all matters relating to a contract modification (change order) during the negotiation of the modification, use of a clause which imposes an obligation on the contractor to submit a price breakdown required to "cover all work involved in the modification" cannot be used to deprive the contractor from his right to file claims.

Conflicts in Plans and Specifications

This is an often misunderstood area in contractor claims. The probability of recovery as the direct result of such conflicts is good insofar as the settlement is limited to the cost difference between the project cost as the plans and specifications are interpreted by the owner or architect-engineer and the contractor.

Public works contracts are called contracts of adhesion, which is a term applied to contract documents that are drawn by one party and offered to the other party on a take-it-or-leave-it basis, where there can be no discussion of terms or contract modifications by the other party. The contractor, however, does have one advantage. In the case of ambiguity, the court will interpret the contract in the contractor's favor. This does not relieve the contractor from the obligation of building the work in accordance with the interpretation of the architect or engineer, but only assures that the contractor will get aid for its trouble.

Frequently, the contractor will find in the specifications that outdated standards are specified, or that products are named that no longer exist. Often, too, the specifications will contain references that wherever codes or commercial standards are specified, you are obligated to use the latest issue of that standard existing at the time the project went to bid. Unfortunately, the designer may never have stopped to think that the design was based upon an old standard that existed in its files or a standard that was only current during the design phase. Occasionally, serious difficulties arise from such practices, and the contractor certainly should have the right to the project cost difference resulting from this kind of error.

Perfect specifications are hard to find. In fact, there has never been one, so the contractor should make a reasonable interpretation at the time of bidding a job, and will then be in a good position for recovery if a variance exists. The contractor's interpretation should be based upon what a reasonable person would interpret the documents to mean; then, the contractor will have the court on its side.

As a protection to itself, a contractor should never attempt to construct any questionable area without first submitting to the owner or architect-engineer for clarification or notifying them of an error. In many contract forms, failure to do this may serve as a bar to full recovery of contractor costs.

Miscellaneous Problems

Problems may arise that are directly related to the conduct of the owner's representative on the job, such as issuing changes in the work that are of such magnitude as to constitute a cardinal change (creating a breach of contract) or indirectly related as in the case of a contractor that may negligently delay another, resulting possibly in the owner seeking recovery from the negligent contractor in order to pay the contractor that was harmed.

Some of the types of problems that may fall into this category include the following:

Damage to work by other primes.
Breach of contract.
Cardinal changes.
Work beyond contract scope.
Beneficial Use of the entire project before completion.
Partial Utilization of the project before completion.
Agency nondisclosure of site-related information.
Agency's failure to make payment when due.

SUBCONTRACTOR RELATIONS

Dispute Clauses

No other set of provisions needs to be coordinated between the prime contract and the subcontracts as much as those relating to dispute resolution. Whether they deal with disputes that arise during the course of the project or after the project has been completed, a lack of coordination between contracts can have disastrous results for the contractor.

Procedural Requirements

Almost every prime contract contains provisions setting forth the procedural requirements for dealing with disputes that arise during the project between the agency and the contractor. When the subcontracts are drafted, the procedural requirements contained in the prime contract must be taken into account. If the prime contract sets forth a *notice* requirement and a procedure for the contractor to follow to assert claims against the agency, then the subcontracts must contain provisions that will enable the general contractor to comply with those requirements should a need for a claim arise. If the prime contract requires 10 days' notice of a claim from the contractor, then a subcontract requiring 20 days' notice of a claim will be of little value to the contractor. If the prime contract sets up the architect-engineer as the initial arbitrator of disputes, the subcontracts must be drafted to take that requirement into account. Whatever the procedural requirements, the subcontracts must enable the contractor to meet them.

Arbitration vs. Litigation

Whether a person prefers binding arbitration or litigation in court usually depends upon one's past experience and who is holding the money. If a contractor has been successful in arbitration but not successful in court, the contractor will probably swear by arbitration. On the other hand, a contractor that believes it was "burned" by arbitration will probably head for the courtroom. Generally, binding arbitration before arbitrators knowledgeable in the construction industry will be cheaper and faster than court proceedings in all but the largest and most complicated disputes. For that reason, the party seeking recovery usually favors arbitration, whereas the party holding the money wants to keep it and usually favors the courts.

Regardless of whether the prime contract specifies arbitration or leaves litigation to the courts, subcontracts must be drafted with whatever is specified in the prime contract. The contractor must carefully consider what is in its best interests under the circumstances and draft the subcontracts accordingly. Consistency is the safest course. If the prime contract contains an arbitration clause, the subcontracts should contain the same clause, and the subcontractors made parties to and bound by any arbitration award. If the prime contract leaves disputes to the courts, the subcontracts should do likewise. By bringing the subcontracts dispute provisions in line with those contained in the prime contract, in all but the most special situations, the contractor can avoid being subject to multiple actions in multiple forums.

To illustrate our point, suppose a contractor enters into a prime contract that does not contain an arbitration clause. The contractor then contracts with its subcontractors using a standard subcontract containing an arbitration clause. A dispute arises regarding a subcontractor's performance and the agency withholds payment to the contractor because of it. Having not been paid, the contractor does not pay the subcontractor. The subcontractor then commences arbitration against the contractor seeking payment. The contractor, due to the lack of coordination between the terms of the prime contract and the subcontract, is caught in the middle and must now file a court action against the agency for payment if the dispute cannot be settled by negotiation.

In the arbitration, which would probably take place before the trial with the agency, the contractor will have to maintain that the subcontractor is not entitled to payment due to nonperformance. In court, the contractor will have to take the opposite position and maintain that the subcontractor is entitled to payment. Worse yet, the contractor might have to pay an arbitration award out of its own pocket without any assurance that it will prevail in its court action with the agency and ultimately get paid! The problem would not have occurred had the subcontract not contained an arbitration clause.

The award of attorney's fees and costs is a dispute clause that must be coordinated in the subcontracts as well. Attorney's fees and costs can represent a major portion of any claim or dispute, and the possibility of recovering these costs can impact heavily on whether the disputes are settled or fully litigated in court or in arbitration. As in the example just cited, the subcontract must contain a well thought out award of attorney's fees and costs provision that is coordinated with the prime contract, or the award of attorney's fees and costs may be available in one legal action and not in another.

Damages for Delay

When a prime contract contains language restricting the contractor's right to seek damages for delays suffered on the project, the subcontracts should be coordinated to reflect the same limitations. Without coordination in the subcontracts, a subcontractor might be able to recover more from the contractor by way of damages for delay than the contractor could recover from the agency.

CHAPTER 11   CONSTRUCTION CLAIMS

Unforeseen or Unusual Site Conditions

Unforeseen site conditions are probably responsible for more construction claims and disputes than any other condition. As a result, prime contract documents very often contain provisions that address the issue of unforeseen site conditions or unusual site conditions, and how these will be handled under the contract (see "Differing Site Conditions" earlier in this chapter). Any subcontract drafted for use with such a prime contract should coordinate with the prime contract and reflect the same terms.

Claims on Subcontractor's Bonds

When subcontractors are required to post a performance bond on a project, either as a part of the prime contract or simply by the general contractor as a way of protecting itself, the contractor will have the right to go against that bond should the subcontractor fail to perform its job. Consequently, all notice and claim requirements and procedures required of a party making a claim against a performance bond would be applicable between the contractor and subcontractor. If the contractor sees a claim developing against a bonded subcontractor, proper notice of the claim should be made to the subcontractor and the surety immediately. Then, careful records to back up the contractor's claim must be kept by the contractor.

If subcontractors are required to obtain bonds for the benefit of the contractor, then the contractor should make it a policy to obtain copies of these bonds early on for its records. In that way, the contractor can be aware of any time requirements contained in the bonds for the filing of notices or claims against the bonds. The contractor should also be familiar with any statutory requirements relating to claims on such bonds in the jurisdiction where the project is being constructed.

DOCUMENTATION FOR CLAIMS

Adequate documentation is the key to sustaining a successful claim. Keep detailed, written records of every contact with the owner, architect-engineer, subcontractors, suppliers, and their representatives. At an early stage of the project, it is desirable to organize the filing system so that documents are later recoverable when needed.

Of particular concern should be the keeping of cost isolation data on all matters dealing with delays in the work, extra work ordered, disputed work performed pursuant to field orders issued, work performed on any differing site condition work, and any other work that may be the subject of a later claim. Unless the contractor can prove which individuals were actually performing such work, along with their hours and rates of pay, and unless the contractor can show the actual rental costs of equipment used on specific items of work, as well as the cost of all materials used, its claim may be subject to rejection by a court due to your inability to substantiate the detailed cost of such work. Figure 11.6 illustrates a form useful for such documentation.

The contractor should photograph everything that may help support its position in a claim. When work has been damaged, photograph the damage at the time it was incurred, prior to allowing the subject to be moved or altered in any way.

It is good to remember that the contractor is not the only party that is likely to be filing claims. Public agency claims against the contractor are almost as prevalent as contractor claims. Thorough documentation of all work during the life of the project, whether or not claims are anticipated, is the "ounce of prevention" that may spare a careful contractor significant monetary losses. The camera also plays its part here, as many a contractor laying pipelines in city streets has found it to its benefit to photograph every linear foot of curb and sidewalk frontage along the route of the pipeline prior to even moving onto the site. It is amazing to know how many "honest" people there are who in all honesty never noticed the crack in their curbs or sidewalks, or the sidewalk that heaved, prior to the contractor's excavating in front of their house. Then naturally, such people honestly believe that this damage was done by the contractor. For pipeline work and other linear projects, a video camera does this job just as well and at a lower cost. Such practices will not only protect the contractor from the average "honest" citizen, but also from those opportunists out there who think that all contractors have deep pockets.

Whenever a contractor is served a notice by the owner or the architect or engineer concerning the quality of its work, for the contractor's own protection it should be photographed in detail. Each time such a photograph is taken, the event should be logged in the field diary along with a description of what was photographed and just what you were attempting to show. This will help verify the date and time that specific photographs were taken.

E. R. FISK CONSTRUCTION

DAILY EXTRA WORK REPORT

PROJECT TITLE PARKHURST LIFT STATION
PROJECT NO. R-407 LOCATION CITY OF MERCER
WORK PERFORMED BY E.R. FISK CONST.
DESCRIPTION OF WORK HAUL & ERECT CONSTRUCTION SIGNS

DATE PERFORMED 6 AUG 1987
DATE OF REPORT 10 AUG 1987

REPORT NO. 49
AMOUNT AUTHORIZED $
PREVIOUS EXPENDITURE $ -0-
TODAY $
TO DATE $
CONTRACTOR JOB NO. 87-0157
CONTRACTOR REPORT NO.

Equip. No.	Equipment	Hours	Hourly Rate	Extended Amounts
222	FLATRACK 19,500 GVW	8 (10)	22.80	228.00
111	PICKUP UNDER 12,000 GVW	6 (10)	13.05	130.50
R-1	HAND-HELD AUGER UNDER 5-HP	1-DAY	per day	6.65

Data used are fictitious for illustration only

MATERIAL and/or WORK done by specialists

Description	No. Unit	Unit Cost	
LUMBER & HARDWARE See Attached Invoice #21704		225.55	225.55

TOTAL COST OF EQUIPMENT, MATERIALS AND WORK

P.R. No.	Labor	Hours		Hourly Rate	Extended Amounts
	TRUCK DRIVER PAUL MILLIGAN	O.T.			
		Reg.	8	17.25	138.00
	LABOR FOREMAN JOHN RUSSELL	O.T.			
		Reg.	6	16.44	98.64
	LABORER KAHUNA TRAN	O.T.			
		Reg.	8	15.94	127.52
	LABORER GUADALUPE RODRIGUEZ	O.T.			
		Reg.	6	15.94	95.64
	LABORER GERALD MADLEY	O.T.			
		Reg.	6	15.94	95.64
		O.T.			
		Reg.			

SUB TOTAL	555.44
Added Percentage—(See Special Conditions) 16%	88.87
Subsistence ____ No. ____ @$ ____	—
Travel Expense ____ No. ____ @$ ____	—
Other ____	—
TOTAL COST OF LABOR A	644.31
→ B	590.70
+ 20 % ON LABOR COST (A)	128.86
+ 16 % ON EQUIP., MAT'L AND WORK COST (B)	88.61
TOTAL THIS REPORT	1,452.48

R. E. Barnes
Submitted—Contractor

Tom Parker
Recommended—Res. Proj. Representative

DISTRIBUTION: 1. Proj. Mgr. 2. Field Office 3. File

For Office Use Only
Pd. on est. no. ____
Checked by ____

Wiley-Fisk Form 15-5

FIGURE 11.6 Cost isolation documentation of extra work, disputed work, delayed work, time-and-materials work, differing site conditions work.

RESOLUTION OF CONSTRUCTION DISPUTES

Although the individual provisions may vary, most construction contracts for public works contain language regarding the method of resolution of claims, either by litigation or arbitration. In the construction industry, there appears to be far more emphasis on the resolution of disputes by arbitration. Unfortunately, in a contract of adhesion (which includes any public works contract), the contractor is not offered a choice except in those states where arbitration statutes prohibit contract provisions that contain an agreement at the time of signing the contract to submit to binding arbitration on future disputes.

Yet, in many cases, neither arbitration nor litigation is the sensible answer. Keep in mind this: If you go to court or to arbitration, the costs on smaller claims may exceed the settlement even if you win. Whenever the value of a claim is under $20,000 or even $30,000, it is doubtful you can ever win financially. The only true winners will be your lawyers and claims consultants (see Figure 11.7). The soundest advice you can receive is, if you want to stay in business, do not go jousting at windmills. Fighting for principles may be a fine hobby for another Howard Hughes, but for most of us, it is financially unsound. The secret is to negotiate.

Negotiation involves compromise, so enter into it with that in mind. Also, remember that claims in the dollar range previously mentioned cannot ever be won by anyone except the lawyers involved, so even if you collect only ten cents on the dollar, you would actually be money ahead. So put up a

FIGURE 11.7 If your claim gets to court, there will be only one winner—the lawyers.

good fight; do some hard bargaining, but be prepared to compromise. And it should not be forgotten that unless you plan to leave a little on the table for the other party to save face, you may force them into arbitration or litigation, a position which neither of you can really afford.

SUPPORTING THE CONTRACTOR'S POSITION

The organized documentation of the facts surrounding an alleged wrongdoing is the single most important weapon in a case. There is nothing that can replace regular, detailed record keeping during construction. A tremendous volume of documentation is common to all projects, good and bad; therefore, the key is the proper organization of the various documents. It is an important duty of the project manager or project superintendent to provide or assist the attorney in producing a chronologically organized summary of the facts of the dispute. At an early stage, the attorney should obtain copies of documents and other information held by the architect-engineer and the public agency through a pretrial process of discovery before they can be disturbed.

Another important source of evidence is the live witness, including the expert witness. The contractor should provide its attorney with a list of all potential witnesses, including those who may testify against the claim, since it is equally important that the opposition's strengths are known early in the case. A live witness is especially important when the dispute in question concerns an oral agreement, such as an oral change order, for which there was no subsequent written memorandum. Exact spoken words can also provide valuable evidence when a dispute involves some subject matter that was left out of the written memorandum, either by error or because it was not directly relevant to the agreement. Given a solid reputation, the most important qualification of a person who will serve as an expert witness is integrity. The integrity of the expert witness will be tested by the opposition in many ways, but most important, by questioning the witness as to whether or not he or she is too closely connected with the case to give an impartial opinion.

In addition to proving the defendant's liability, the plaintiff is faced with the task of establishing the validity of the damages claimed. The general rule is that the prevailing party may recover whatever damages may reasonably be supposed within the contemplation of the parties at the time that the contract was made. This is the most complex and time-consuming phase of construction litigation. Often, the case is split into two separate trials, one for determining liability and the other for determining the damages. The advantages of separating the trial into two separate actions are that the determination of liability issues may narrow and simplify preparation for the damages phase; it may enable the parties to get an earlier determination of the entire dispute by settlement after the liability issued has been decided; and it may reduce the number of parties involved in the damages phase. The greatest disadvantage is that separation into two separate actions may delay final resolution.

Burden of Proof

Generally, it is the plaintiff that has the burden to prove its claim. When the contractor succeeds in proving substantial compliance against the public agency's claim of nonperformance, the burden of proof reverts to the public agency to prove its claim for residual defects.

Method of Presentation

The ordinary fact finder, whether it be a jury panel or a judge, is not likely to be familiar with the technical and administrative details of construction. Therefore, the factual issues must be presented in a simple, concise manner. Photographic evidence and charts are very effective in conveying the facts. Actual site visits may save time and energy consumed in otherwise describing the general physical conditions. An expert witness who can translate technical concepts into simple language is an invaluable evidentiary aid.

Evidence

In all disputes, regardless of the method of resolution, the prevailing party will be the one that can better support the burden of proof of its allegation. In preparing to bring a dispute to some kind of resolution, the contractor must assure that it is equipped with all of the necessary evidence in support of its claim or defense and anticipate what its opponent will produce as evidence of its position on the claim. The following is a checklist of the possible sources of evidence that should be used in the course of a dispute settlement:

Documentation

1. Bid documents.
2. Boring logs and soils reports.
3. Drawings: as planned.

CHAPTER 11 CONSTRUCTION CLAIMS

    4 Drawings: as built.
    5 Specifications.
    6 General and Supplemental General Conditions.
    7 Schedule: as planned (CPM, etc.).
    8 Schedule: as built (progress charts).
    9 Addenda.
  10 Change Orders.
  11 Instructions and directives.
  12 Inspection records.
  13 Contractor's logs.
  14 A&E or agency's Diaries (through deposition).
  15 Correspondence files.
  16 Check registers.
  17 Purchase orders.
  18 Shipping and delivery tickets.
  19 Time cards.
  20 Man-loading charts.
  21 Memoranda (including memos to file).
  22 Site photographs.
  23 Testing results.
  24 Public records.

Witnesses

    1 Contractor's project personnel.
    2 Public agency personnel.
    3 Architect-engineer personnel.
    4 Construction manager personnel.
    5 Subcontractors.
    6 Inspectors.
    7 Suppliers.
    8 Testing lab personnel.
    9 Consultants (expert witnesses).
  10 Adverse witnesses.

CONSTRUCTION CLAIMS RESOLUTION ALTERNATIVES

TIME	SETTLEMENT COST	BINDING NATURE	APPEAL
Negotiation 1. Dependent on the parties' negotiators' objectives, attitude, and other factors. 2. Can be very fast.	1. Minimal. 2. Cost of compromised settlement.	1. Take it or leave it. 2. May lead to an agreement.	1. Waived if agreement is reached. 2. Arbitration or litigation if no agreement.
Mediation 1. Same as negotiation. 2. There may be some limitations imposed by the mediator's schedule. 3. Usually fast.	1. Mediator's compensation, if any.	1. Take it or leave it. 2. May lead to an agreement. 3. Moral pressure to reach an agreement.	1. (Same as above.)
Mediation–Arbitration 1. Speed depends on the procedure used. 2. If formalities can be waived, resolution is fast.	1. Mediator's compensation, if any.	1. May be agreed in advance in most states that parties will be bound to the decision.	1. (Same as above.)
Arbitration 1. Faster than litigation. 2. Rules may impose some limitations. 3. Availability and schedule of arbitrators is a problem. 4. Preparation may take several months.	1. Filing fee. 2. Arbitrator's compensation after second day. 3. Attorney's fees if any.	1. May be nonbinding or binding according to contract.	1. No review of merits in court. 2. Arbitrator not required to explain award.

FIGURE 11.8 Construction claims resolution alternatives.

CONSTRUCTION CLAIMS RESOLUTION ALTERNATIVES			
TIME	SETTLEMENT COST	BINDING NATURE	APPEAL
Litigation 1. May take up to five years or more to reach a trial. 2. Preparation itself may take years.	1. Prohibitively expensive, both in terms of attorney's fee and time costs.	1. Binding.	1. Full appeal.
Drop claim/concede 1. None.	1. Value of claim.	1. Contractual agreement by mutual accord. 2. Waivers.	1. None, right waived in most cases.

FIGURE 11.8 (*Continued*)

There is no claims protection or support without adequate documentation. The single most important thing that a contractor can do prior to the existence of a claim is to provide its attorney with the tools needed for a contractor to defend or prosecute its case, whether it be in arbitration or litigation.

ALTERNATIVE METHODS FOR DISPUTE SETTLEMENT

Figure 11.8 is a simplified summary of the principal methods available to the contractor for the resolution of construction disputes.

CHAPTER 12

Project Closeout

- Completion of the Work
- Punch List Procedures
- Cleanup of Project Site
- Final Submittals
- Beneficial Use or Partial Utilization
- Substantial Completion
- Final Payment
- Release of Retainage
- Liquidated Damages for Delay
- Liens and Stop-Payment Orders
- Pickup Work

CHAPTER 12 PROJECT CLOSEOUT

COMPLETION OF THE WORK

At first, it would seem that all that would be necessary to close out a project would be to have the architect-engineer inspect and accept it for the agency, then submit the final payment request and move off the jobsite. But first, there are a number of other considerations in the more formalized area of public works construction.

PUNCH LIST PROCEDURES

The primary purpose of the recommended punch list procedures is threefold.

1. To assure the public agency and the architect-engineer that all work on the project is completed on a timely basis in accordance with the provisions of the Contract Documents.
2. To provide the contractor and the subcontractors with positive incentive to perform their work in a manner that they believe will lead to prompt and full payment.
3. To prevent the issuance of multiple punch lists, which frequently cause delays in job completion and prevent clear-cut goals in completing a project.

Work by the Contractor and Subcontractors

The contractor and each subcontractor should carefully and regularly check their work while it is being performed. The contractor's superintendent should prepare and maintain a written record of all observed deficiencies as the job progresses so that remedial action may be taken promptly and no deficiencies forgotten or overlooked. Unsatisfactory work should be corrected immediately and should not be permitted to remain and become a part of the punch list. Corrections of each subcontractor's work should be made before they are allowed to leave the project.

During the finishing stages of each subcontractor's work, the contractor and the architect-engineer, accompanied by the subcontractor, should make frequent and careful inspections of the subcontractor's work, so as to progressively check for and insure the correction of any faulty or deficient work.

Work by the Contractor

Work performed directly by the contractor with its own forces, and the final acceptance of such work and the overall project, would normally be covered by and accepted according to the same procedures outlined for work performed by the subcontractors.

If the agency, or an installer of agency-owned equipment or furnishings, damages work by the contractor that has been satisfactorily completed and accepted, the architect-engineer is under an obligation to promptly advise the agency of its obligation to repair such damaged work. If the agency wants the contractor to make the repairs or to perform maintenance work that is not called for in the Contract Documents, the contractor is entitled to separate reimbursement for this activity.

Time of Prefinal Inspection

Ideally, when the contractor determines that the work of a subcontractor (or a group of subcontractors doing integrally related work) has been completed satisfactorily and in accordance with the terms of the Contract Documents, the architect-engineer should be promptly notified.

Upon such notification by the contractor, the architect-engineer should make arrangements for the prefinal inspection of that portion of the work. Representatives of the contractor and the subcontractors should be prepared to answer questions that may be posed by the architect-engineer.

Typically, however, the public agency or its architect-engineer prefers to limit this level of inspection to the end of the project, thus creating the attendant delays, multiple punch lists, and last-minute corrections of details that could have been cleaned up more efficiently and economically had the foregoing recommendations been followed.

Preferably prior to, but no later than during, this prefinal inspection period, dates should be established for equipment testing, acceptance periods, warranty dates, and instructional requirements not previously completed or agreed on.

Preparation of the Punch List

Following the prefinal inspection of the contractor's work, the architect-engineer should prepare a *Punch List* (see Figure 12.1), setting forth in accurate detail any items of work that are found not to be in accordance with the requirements of the Contract Documents. Following preparation of the

PUNCH LIST

Project: Zone 3 Booster Pumping Station   No. 532-88
Location: Mission Viejo, Orange County, California   Date 8 December 1987
Inspection was conducted at above project by R. M. Bendix at 2:30pm o'clock this date.

REPRESENTATION

CONTRACTOR-OWNER
- Contractor: Farris & Booth, Inc.
- Owner: Santa Ana Water District
- Improvement District No. 2-E

ENGINEER-ARCHITECT
- Design: VTN Consolidated, Inc.
- Staff Spec.: C. H. Lawrance, P.E.
- Job Supervisor: R. M. Bendix

The following items are to be corrected or completed to comply with the contract documents:

Type of Inspection	Check	Final XX	1 Yr. Guar.	Guar.

NO.	ITEM
9	Cracked globe on vapor-proof fixture in lower gallery
13	Field coating req'd for damaged shop paint on LPG tank
17	Replace defective stair tread
18	Replace door No. 2 Stop W/331ES or 431ES
19	Finish paint access panels over factory prime
20	Readjust swing of door No. 5
21	Master-keying not in accordance w/spec
28	Readjust air systems to within 5% of design requirements
31	Provide handwheel for emergency operation of plug valve motor operator as per spec.
37	Adjust access MH cover to finish grade
39	Replace existing welded gate post with specified one-piece hot-dip galvanized after fabrication post
40	Clean up premises — Remove all debris from the site
41	Remove temporary power facilities
43	Re-test standby engine-generator set on automatic mode
44	Pump Station system control check & validation test

Data used are fictitious for illustration only

[Signed] R. M. Bendix

DISTRIBUTION:
1. Project Manager
2. Contractor's Representative
3. Resident Project Representative
4. File

Wiley-Fisk Form 17-1

FIGURE 12.1 Formalized Punch List form.

Punch List, a meeting betwen the contractor, the subcontractors, and the architect-engineer should be held promptly to discuss any questions concerning what the architect-engineer requires to be done before the work can be accepted as complete.

Final Punch List Inspection

When advised by the contractor that the Punch List items have been completed (see Figure 12.2), the architect-engineer, accompanied by the contractor and the subcontractors, should conduct the final inspection of the various subcontractors' work and should then, if the Punch List items have been completed satisfactorily, issue a certificate for final payment, less the specified retainage, for that portion of the work accepted by the public agency.

If, following the final inspection of a subcontractor's work, there remains a question as to whether one or more Punch List items have not been properly completed, but the overall work has been substantially completed (see the definition of "Substantial Completion" later in this chapter), the architect-engineer should issue a certificate that notes the uncompleted Punch List items.

CLEANUP OF PROJECT SITE

Normally, all public works specifications require that the jobsite be kept clean by the contractor during the progress of the work. The contractor is similarly obligated to *thoroughly* clean up the construction site at the end of the job before the project can be accepted. The final cleanup is of significantly greater proportions than previous cleanup work, as all of the various items of demobilization rightfully are included under the cleanup category. This includes the removal of all temporary utilities and related facilities, haul roads, temporary fences, field offices, detours, stockpiles, surplus materials, scrap, replacement of landscaping where it had been temporarily removed, street cleaning, and the obtaining of releases from the various city, county, or other governmental authorities having jurisdiction.

The contractor is obligated to clean up the site of its own operations as well as all other areas under the control of the agency that may have been used by the contractor in connection with the work on the project. In addition, the contractor is required to remove all temporary construction, equipment, waste, and surplus material of every nature unless the agency has

CONTRACTOR'S CERTIFICATION OF COMPLETION

TO: Mr. G. H. Brown, P.E.
Director of Public Works
City of Casa Grande
City Hall
Casa Grande, AZ 00000

ATTN: Resident Project Rep.

FROM: B&H Construction, Inc.
(Firm or Corporation)

DATE: 5 May 1989
PROJECT: Arroyo WWTP Addition
JOB NO.:
CONTRACT NO.: 88-0324
OWNER: City of Casa Grande

This is to certify that I, Ray E. Barnes am an authorized official of B&H Construction, Inc. working in the capacity of Project Manager and have been properly authorized by said firm or corporation to sign the following statements pertaining to the subject contract:

I know of my own personal knowledge, and do hereby certify, that the work of the contract described above has been performed, and materials used and installed in every particular, in accordance with, and in conformity to, the contract drawings and specifications.

The contract work is now complete in all parts and requirements, and ready for your final inspection.

I understand that neither the determination by the Engineer-Architect that the work is complete, nor the acceptance thereof by the Owner, shall operate as a bar to claim against the Contractor under the terms of the guarantee provisions of the contract documents.

Data used are fictitious for illustration only

BY: Ray E. Barnes, P.E.
TITLE: Project Manager
FOR: B&H Construction, Inc.

DISTRIBUTION: 1. Proj. Mgr.
2. Field Ofc.
3. File

Wiley-Fisk Form 17-2

FIGURE 12.2 Contractor's written notice of readiness for final inspection.

approved otherwise in writing. Final acceptance of the work may be withheld until the contractor has satisfactorily complied with all of the requirements for final cleanup of the project site. This includes cleanup of city streets as well, where dirt or other deposits have accumulated as a result of the contractor's operations.

Disposal of all waste and refuse is normally required to be at the contractor's expense. No waste or rubbish of any nature will normally be permitted to be buried or otherwise disposed of at the site except upon receipt of written approval of the agency.

FINAL SUBMITTALS

As a part of its closeout obligations, the contractor is obligated to obtain and submit the following items when specifically required under the contract. If such items are not written in the contract, the contractor may be entitled to extra payment for such work:

Guarantees.
Certificates of inspection.
Operating manuals and equipment instructions.
Keying schedule.
Maintenance stock items; spare parts; special tools.
Record drawings (as-built drawings).
Special bonds (roof; maintenance; guarantee; etc.).
Compliance and inspection certificates by local agencies.
Consent of surety for final payment.
Final waivers of lien (varies from state to state).

In addition, where specifically required under the contract, as found on treatment plants and similar complex installations, the contractor may also be required to contract with the manufacturers of the installed equipment to provide on-site training for the agency-operating personnel.

BENEFICIAL USE OR PARTIAL UTILIZATION

Generally, during the administration of contracts involving the construction of treatment plant facilities or similar process plant additions, the public agency will find it necessary to use completed portions of the project well before total project completion. This is generally referred to as beneficial use or partial utilization.

This is a particularly sensitive, high-risk area for both the contractor and the public agency, and several problems can accrue as a result of such utilization:

1. Identification of the causes of latent defects in equipment are almost impossible, and most malfunctions may legitimately be claimed by the contractor to be the result of improper maintenance by the public agency.
2. Expiration dates of equipment warranties and the term of the contractor's guarantee period are in question.
3. Maintenance responsibility frequently becomes a controversial issue between the contractor and the public agency.
4. Security of the site and on-site safety responsibility are no longer a clear-cut issue during beneficial use. Generally, the public agency will inherit the responsibility for these, thus relieving the contractor of this risk.
5. One of the principal risks involved in Items 3 and 4 is that each party will assume that the other is responsible for maintenance, security, and safety, and neither party will perform these tasks. This can result in increasing the severity of the problem and increasing the volume of disputes and claims.

It is recommended that whenever practical, the contractor should advise the public agency of the risk of beneficial use or partial utilization and make it clear that the public agency must accept all attendant risks. The public agency should be advised by the contractor of any delaying effects of beneficial use or occupancy by the agency's forces, and the fact that any added costs of delay will have to be paid by the public agency.

If the public agency does elect to utilize any portion or portions of a project prior to total project completion, the following guidelines are recommended:

1. Issue a letter to the agency that advises it that the taking of the beneficial use of a particular described portion of the work as of a specified date will serve to begin the product warranty date, if it has not already begun to run, and that as of then the agency must assume certain responsibilities.

2. A copy of the letter should be sent to the public agency's operations and maintenance department.

3. Advise the public agency to take positive steps to assume the maintenance of any affected equipment.

4. Advise the public agency that they will then be responsible for security measures, as necessary, to protect the equipment in use.

5. Unless the terms of partial utilization or beneficial use have been clearly spelled out in the contract documents, the contractor should not permit use or occupancy by the forces of the public agency until such agreements have been formally reached.

SUBSTANTIAL COMPLETION

Among lay people, there seems to be a great amount of confusion as to the meaning of the term *substantial completion* of a construction project. The term substantial completion has the same meaning as *substantial performance,* and the definition of substantial performance, quoted from Black's Law Dictionary, Revised Fourth Edition, West Publishing Company, 1968, follows:

> *Substantial performance* exists where there has been no willful departure from the terms of the contract, and no omission in essential points, and the contract has been honestly and faithfully performed in its material and substantial particulars, and the only variance from the strict and literal performance consists of technical or unimportant omissions or defects. [*Cotherman v. Oriental Oil Co.,* Tex. Civ. App. 272 S.W. 616, 619; *Brown v. Aguilar* 202 Cal. 143, 259 P. 735, 737; *Cramer v. Esswein,* 220 App. Div. 10, 220 N.Y.S. 634; *Connell v. Higgins,* 170 Cal. 541, 150 P. 769, 774.] Performance except as to insubstantial omissions with compensation therefor [*Cassino v. Yacevich,* 261 App. Div. 685, 27 N.Y.S. 2d 95, 97, 99.]

The general rule is that the substantial performance of a contract will "support" the recovery of the contract price, less allowances for defects in performance or damages for failure to comply with the contract strictly. Substantial performance can likewise prevent the termination of a contract for default. If the government can make beneficial use of a substantially completed item, then the contract may not be defaulted. The contractor would be entitled to payment, less deductions for delay and final completion. This

LAWRANCE, FISK & McFARLAND, INC.
CONSULTING ENGINEERS • SANTA BARBARA • ORANGE

CERTIFICATE OF SUBSTANTIAL COMPLETION

TO: City of Lompoc, California — OWNER

DATE OF SUBSTANTIAL COMPLETION: October 12, 1988

PROJECT OR SPECIFIED PART SHALL INCLUDE: Total facility

PROJECT TITLE: Emerson Reservoir
PROJECT NO.: 743-32
LOCATION: Lompoc, CA
OWNER: City of Lompoc, CA
CONTRACTOR: E. R. Fisk Construction
CONTRACT FOR: Reservoir & Pumping Plant
CONTRACT DATE: July 14, 1987

The Work performed under this contract has been inspected by authorized representatives of the Owner, Contractor, and Engineer-Architect, and the Project (or specified part of the Project, as indicated above) is hereby declared to be substantially completed on the above date.

DEFINITION OF SUBSTANTIAL COMPLETION
The date of substantial completion of a project or specified area of a project is the date when the construction is sufficiently completed, in accordance with the contract documents, as modified by any change orders agreed to by the parties, so that the Owner can occupy or utilize the project or specified area of the project for the use for which it was intended.

A tentative list of items to be completed or corrected is appended hereto. This list may not be exhaustive, and the failure to include an item on it does not alter the responsibility of the Contractor to complete all the Work in accordance with the contract documents.

LFM, Inc.
ENGINEER-ARCHITECT

BY: *[signed]* G.L. McFarland — 8-10-88
AUTHORIZED REPRESENTATIVE — DATE

[Stamp: Data used are fictitious for illustration only]

The Contractor accepts the above Certificate of Substantial Completion and agrees to complete and correct the items on the tentative list within the time indicated.

E. R. Fisk Construction
CONTRACTOR

BY: *[signed]* E. R. Fisk — 8-15-88
AUTHORIZED REPRESENTATIVE — DATE

The Owner accepts the project or specified area of the project as substantially complete and will assume full possession of the project or specified area of the project at 10am (time), on Aug 18, 1988 (date). The responsibility for heat, utilities, security, and insurance under the contract documents shall be as set forth under "Remarks" below.

City of Lompoc, California
OWNER

BY: *[signed]* Director of Public Works — 8-12-88
AUTHORIZED REPRESENTATIVE — DATE

REMARKS: (Attach additional sheet, if necessary)
Wiley-Fisk Form 17-3

Telemetering system not yet operable. Delays in receiving components. Entire system to be operable within 30 days of above date. Other pickup work; See attached punch list.

FIGURE 12.3 Example of a Certificate of Substantial Completion.

principle of law is generally applied to construction contracts. [See ENG BCA No. 3610 (June 29, 1984)]

As the contractor, then, you must realize that although substantial completion will generally relieve you of liability at the site as well as any liquidated damages that may be assessed, it also terminates the contractor's insurance policy and begins the warranty and guarantee periods. Special note should be made of outstanding punch-list items, however. The agency has the right to expect these to be completed in a reasonable amount of time . . . like a month or so, and failure to do this may result in your inability to collect the full amount of the contract. The Associated General Contractors of America and the Construction Industry Affairs Committee of Chicago, in their publications, advise the owner to withhold from the money otherwise due to the contractor an amount substantially greater than the value of the uncompleted punch list items. Failure of the contractor to complete these items will cause a forfeiture of the amount withheld to assure completion of the work, and the agency may use the money withheld to get someone else to complete the unfinished work. On the other hand, if a contractor just walks off and is willing to forfeit the amount withheld, there is legal precedent for the contract to be considered substantially complete, and thus the contractor cannot be declared in default.

Read the contract provisions carefully. The term "substantial completion" may be redefined in that document, thus requiring you to reevaluate your position when the time arrives for final payment for the work.

Typically, on public works contracts, a formal notice of substantial completion is executed after *acceptance of the work* by the agency (see Figure 12.3). In some states, it is mandated by law, and the document must be recorded with the county recorder. On public works contracts, unlike private works, "completion" is established by the *acceptance* of the legislative body of the public agency, not necessarily by the actual completion of the work.

FINAL PAYMENT

Withheld Funds from Final Payment

It is not uncommon for the final payment to be made after the deduction of a specified retainage, and less an additional amount that the architect-engineer reasonably estimates would be required to cover more than the cost of completing any Punch List items. This additional amount would be held only until the satisfactory completion of the remaining items, if completed within a reasonable length of time, and possibly forfeited entirely if the pickup item of work is not completed. The Construction Industry Affairs Committee of Chicago suggests that the architect-engineer withhold two times the value of any remaining punch list items.

On private contracts, the retainage can be made payable at the time of the final payment; however, on public works contracts, sometimes the law requires that the public agency hold the retainage until the end of the lien (stop notice) filing period established by statute.

Neither the Punch List nor the final acceptance of any portion of the work should be used arbitrarily to establish commencement dates or warranties or guarantees. Such dates are established through contractual agreements in the Contract Documents.

Nothing in the recommended Punch List procedures should be interpreted as relieving the contractor of its contractual obligations.

Final Payment and Waivers of Liability

How many times has a contractor seen the clause in the General Conditions of the Contract that states the final payment terminates the liability of the agency and that acceptance of the final payment check constitutes a waiver of all claims? Too many times, of course. If you happen to be constructing projects in California, however, the problem is less pervasive, as the legislature in 1982 enacted a law that made such clauses null and void. [Public Contract Code §7100].

Elsewhere, however, you may have to take alternative measures. One method is to divide your payment requests so that the "final payment" requested is only a few hundred or thousand dollars, and then you can afford to accept the larger payment without waiving your rights. The smaller uncollected payment will simply be your investment to keep the account open. Just *do not bill the agency for the final payment*. It will drive them up the wall and send their bean counters into a frenzy, but it's an old method used on federal contracts in the past.

What you really need is an agreement with the agency that the payment of undisputed amounts will not act as a release of all claims against the public agency arising by virtue of the public works contract related to those amounts. Before you bill for and cash any check for the alleged final payment, be certain that you obtain a written agreement from the public agency specifically excluding disputed contract claims from the operation of the release. This, of course, applies to claims filed prior to final payment. You still must expect to waive your rights to file claims after final payment.

RELEASE OF RETAINAGE

The primary purpose in withholding retainage from the contractor is to assure the satisfactory completion of a project. On public works contracts, agencies are often required by law to withhold the return of retainage as long as there are outstanding lien claims (stop notices filed). However, there is a way out for the contractor if this occurs. That is, it can obtain a Stop Notice Release Bond from a surety different from the one that wrote the payment and performance bonds. Upon the presentation of this bond to the agency, it is relieved of all liability under the stop notices and is authorized to release all retainage funds unless there are still outstanding Punch List items which are sufficient to bar declaring the project to be substantially complete.

A contractor cannot expect to receive the retainage funds at the time of substantial completion, as some seem to believe. In most cases, such retainage funds must be held by the public agency until the end of the statutory lien-filing period for the jurisdiction in which the project was built. The terms of the retainage-holding time should be spelled out in the Contract Documents, however, and if this matter is not settled when the documents are prepared, an agreement during the preconstruction conference should be established.

In a few cases, some public agencies have actually held retainage as a club over the head of the contractor as a form of economic blackmail as long as there were unsettled contractor claims against the agency, pending the outcome of litigation, which could take years. This is a matter for your attorney to become involved with, as no agency has legal authority to withhold retainage for this purpose. It may, however, continue to be held as a hedge against stop-notice claims filed by subcontractors and suppliers that allege nonpayment by the general contractor.

LIQUIDATED DAMAGES FOR DELAY

Most construction contracts now contain a liquidated damages provision that states a fixed sum should be paid by the contractor for each day's delay in the completion of a project beyond the agreed upon completion date. Sometimes, a liquidated damages provision relates to the completion of a particular part of a project by a certain date. The liquidated damages amount is most frequently in the amount of $50 to $1000 per calendar day, or more, depending upon whether the particular circumstances set forth in the contract are a reasonable estimate of what damages the agency will suffer if the project is not completed on time. Generally, as long as the liquidated damage amount is not deemed to be a penalty, the courts will honor a liquidated damage provision. If the amount of the liquidated damages is not unreasonably disproportionate to the actual damages suffered, the liquidated damages provision will not be considered a penalty. Sometimes, the courts look to whether or not time is of the essence in a contract. When it is not specifically stated that "time is of the essence," the agency may not be in a strong position to assess liquidated damages. [*Kingery Construction Co. v. Scherbarth Welding, Inc.,* 185 N.W. 2d 857 (1971)]

When a contractor does not complete the project by the projected completion date, the agency will normally give notice that it will be assessing liquidated damages. It will then either withhold the amounts of liquidated damages accruing from each progress payment until completion and acceptance of the work, or will deduct the total amount of liquidated damages from the sum due the contractor at the time of final payment. Obviously, if there is a dispute over the cause of the delay that resulted in the assessment of liquidated damages, then the agency's withholding from the progress payments the amount of the accrued liquidated damages can impact heavily upon the contractor's cash flow.

Cause of Delay

The cause of the delay that led to the assessment of liquidated damages is important in determining whether or not the assessment of such liquidated damages was proper. Some courts have held that an agency-caused delay completely discharges the contractor from the liquidated damages provision, whereas others have held that it merely extends proportionately the time allowed for completion. The prevailing rule is that when an agency has caused a substantial delay in the progress of the work, the time limit set forth in the contract and any provision for liquidated damages based upon that time are unenforceable, and the contractor is responsible only for the completion of the project within a reasonable time.

However, once a contractor has been relieved of liquidated damages as a result of an agency-caused delay, it is not entitled to disregard the time considerations altogether. The contractor must proceed to complete the work in a reasonable time, and any unreasonable delay caused by the contractor may then be the basis for assessing liquidated damages at the rate specified in the contract.

Contractor's Rights

The inclusion of a liquidated damages provision in a contract does not limit or preclude the contractor's right to extensions of time, recovery of damages, or both against an agency for causing them. Where there is an agency-caused delay, the contractor may be entitled to compensation for such items as idle equipment, management and supervisory costs, interest on borrowings, increased labor and material costs, increased bond premiums, additional manpower, impact costs, and other similar items. It should be noted that an agency-caused delay includes any delay caused by any of the agents or consultants that were retained by or worked on the project for the agency.

In a few jurisdictions, courts have apportioned liquidated damages between the contractor and the agency even when the contract did not provide for such apportionment. In such cases, the damages for contractor-caused delay were set off against the damages resulting from an agency-caused delay.

Liquidated Damages Are Not a Penalty

Liquidated damages are sometimes viewed by contractors as a penalty for failing to achieve the impossible. Based upon the rather haphazard method that many architect-engineers use to estimate the length of construction contract time for a contract, that may be a more accurate definition than the architect-engineer would like to believe. However, public contracts usually contain a clause that states that the contract must be completed within a specified time, and that the contractor will be obligated to pay liquidated damages for every day beyond the specified completion date that the project remains unfinished.

The usual clause states that it is agreed by contract that the public agency will be financially damaged if the work is not completed on time, and that it would be either impossible or extremely difficult to determine the exact amount of the damages. Therefore, the contract provision goes on to say that the parties to the contract have in good faith attempted to determine the extent of the damages that would be suffered by a delay in completing the work, and that therefore, the agreed on *liquidated damages* would be fixed at some set dollar amount. This value, then, will take the place of actual damages in the case of a delay.

Liquidated Damage Clauses in Subcontracts

If a contractor can be assessed liquidated damages on a daily basis if the job is not completed on time due to the failure of a subcontractor to perform, the contractor must eventually take that into consideration when drafting its subcontracts. If the contractor is assessed liquidated damages by the agency, the terms of the subcontracts should enable the contractor to pass this cost along to the responsible subcontractor. The assessment of liquidated damages is a very serious matter. For, if they are assessed over a long enough period of time, the completion of the project becomes an economic impossibility.

Amount of Liquidated Damages

In most cases, such damages are set by engineers of the public agency based upon old specifications. Thus, the contractor will find that the amounts are usually far less than the actual cost to the agency if the project was actually delayed. In fact, even today, it is not uncommon to see public contracts with liquidated damage clauses in the amount of $200 or $300 per day! As a contractor, you can immediately see the rather advantageous position that this places you in. After all, if a project is delayed, wouldn't any contractor rather ante up $200 or $300 per day in liquidated damages rather than pay the considerable higher cost of accelerating the work?

"Time is of the Essence"

Before a contractor gets tripped-up on a liquidated damages provision, however, it should look in the contract for the words "time is of the essence." This is the key to the enforcement of such a clause by the public agency. Case law supports the concept that in the absence of the phrase, "time is of the essence," the specified completion date would be interpreted by the court as a target date only, and that the absolute date of completion originally specified is unenforceable.

LIENS AND STOP-PAYMENT ORDERS

Mechanic's lien laws applicable to construction are designed to protect subcontractors, material suppliers, laborers, and in some cases architects, engineers, and other design professionals who contribute to a work of improvement. These individuals and entities are all potential lien claimants on the projects they worked on. Because of the wide variation in state laws, it is impossible to do more than discuss lien rights in a very general way.

Although public property is not subject to liens, some states have lien laws that entitle an unpaid claimant to place a lien on public funds that may be due a contractor. Under such schemes, the unpaid claimant advises the agency of its claim in accordance with the statutory notice requirements, and then the agency must stop further disbursements to the contractor of the affected funds until payment has been received by the claimants. Generally, once a claim has been made, the lien claimant must foreclose on its lien within a set time period in order to collect; otherwise, its rights will be lost. The agency must hold the liened funds until the lien has been satisfied through foreclosure, at which time the funds are paid over to the claimant. California, Louisiana, New York, and Texas are examples of four states that have provisions allowing liens to be filed against construction funds. In California, a lien against construction funds is accomplished by the use of a *"stop notice,"* which, when served, gives the agency notice to stop payment to a contractor.

A contractor should make every effort to ensure that timely and correct payments are made to every entity working on a project, as the potential disruption of the cash flow that can result from the filing of such liens against funds due to the contractor can have devastating results. If subcontractors, material suppliers, sub-subcontractors, and the like are entitled to file such liens, a contractor must work to minimize and avert claim situations that could lead to such liens. A contractor must set up a release and lien waiver system that will help ensure that money paid out by the contractor gets into the right hands.

Contractor Protection

Most states that have statutory provisions enabling lien claimants to lien construction funds usually have other provisions to help the contractor protect itself against fraudulent, improper, or disputed claims. In some states, if the contractor posts a bond in a specified amount to cover the amount of a lien, the agency is free to then release the construction funds that have been held up in accordance with the original lien. However, it should be noted that agencies will often demand that all liens and lien-related issues be resolved by the contractor before the project is accepted and final payment is made.

Lien Waivers

A contractor may be asked to submit lien waivers as a condition of receiving payment, and although this is not at all uncommon at the end of the job, increasingly partial waivers of lien are being required on a monthly basis, each covering only the value of the currently completed month's work.

There are basically two types of waivers of lien, and two versions of each:

1. Conditional Waiver and Release Upon Receipt of Progress Payment Upon Final Payment
2. Unconditional Waiver and Release Upon Receipt of Progress Payment Upon Final Payment

A contractor will generally be asked to submit lien waivers for itself and each of its subcontractors before an agency will issue a check. Some agencies do not seem to have the least bit of sympathy for the fact that if you issue an unconditional release, it could refuse to pay at all, and the contractor would then have no recourse for collection. The answer, of course, is to either use the unconditional lien waivers when receiving payment in person, trading your lien release for the agency check, or if being paid by mail, to submit only a conditional waiver and release. A contractor is not at risk as long as conditional waivers are used, but do not consent to submitting unconditional releases unless the transaction is being handled in person and a check is tendered at the same time that the release is submitted.

Subcontractor Lien Releases

If the prime contract sets forth a payment procedure that includes the supply of certain releases or lien waivers, the subcontracts should contain those requirements as well. Unless the subcontractor is aware of the need to obtain certain releases or lien waivers from the beginning, major billing problems may result.

PICKUP WORK

Uncompleted Punch List Items

Although a contractor may have received a certificate of substantial completion for a project, it is well to remember that the agency is usually still holding retainage to cover the cost of total completion. The failure to complete all of the items on the Punch List is generally settled by the contractor's forfeiture of that portion of the retainage that represents the cost to the agency of having another contractor complete those items of work, or, an amount that represents the reduced value to the agency of nonconforming or incompleted work.

In terms expressed by the Associated General Contractors of America in one of their publications, it is stated that the owner's architect-engineer "should retain an amount which the architect-engineer reasonably estimates would be required to cover more than the cost of completing any Punch List items," but only until satisfactory completion of the item. The Construction Industry Affairs Committee of Chicago is more specific, suggesting that the amount of such retainage be equal to twice the value of the uncompleted items.

The Certificate of Substantial Completion of the AGC and some others state a completion date for such pickup work. Failure to complete such work is considered as grounds for forfeiture by the contractor, or at least a negotiated settlement amount.

APPENDIX · A ·

Surety and Insurance Bond Checklist

SURETY BOND CHECKLIST

Part I—Types of Bonds

Bid or Proposal Bonds

These bonds provide that if a contract is awarded, the contractor will execute the contract documents and furnish the required performance and payment bonds. The penalties assumed by the surety under bid bonds vary from as little as 2 percent of the total amount of the bid to as much as 20 percent of the total amount of the bid on certain federal projects.

Performance Bonds

In short, these bonds provide for the performance of the agreement by the contractor and payment by the surety of the owner's loss up to the amount of the bond penalty if the contractor defaults.

APPENDIX A SURETY AND INSURANCE BOND CHECKLIST

Labor and Material Payment Bonds

These bonds provide that the furnishers of labor and materials to the project will be promptly paid.

Maintenance Bonds

Maintenance bonds protect the owner, usually for a period of one year, against defects in workmanship and materials. In many states, if the contract provisions are adequate, separate maintenance bonds are not necessary because the performance bond automatically applies to each such maintenance guarantee.

Release of Lien Bonds

These bonds are presented to an owner to obtain payment when a mechanic's or other lien has been filed against the premises or the unpaid contract price. In substance, they assure the owner that he or she will not suffer loss or damage from the lien claimed if, not withstanding this claim of lien, payments continue to be made to the contractor.

Part II—Factors in Bonding

A Surety Bond Is Not Insurance

A surety bond will not insure the owner or the contractor against loss or guarantee against financial failure. Basically, it is comparable to a conditional promissory note guaranteeing protection to the owner or obligee.

Considerations When Bonding

Surety companies consider many factors before underwriting a contract bond. When requesting bonds, a contractor should be prepared to present factual information regarding its firm, as well as information on:

1. Recent projects constructed
2. Financial status, including bank credit and secondary assets
3. Equipment and machinery, including amount immediately available
4. Organization, including experience, skill, and responsibility of key personnel

Select Surety Carefully

To facilitate the handling of bonds, contractors should select a thoroughly qualified and experienced agent who has legitimate surety company affiliations to enable him or her to speak for and bind the company represented and to offer promptly the bonding service needed by the contractor. Obtain from the bonding agent a certified copy of the power of attorney that clearly sets forth the extent of his or her authority to act for the surety company.

Forced Placement of Insurance and Bonds

Any action or contractual requirement depriving the contractor from selecting a reputable surety company of its own choosing is not only a departure from traditional practice and contrary to the best interests of both the contractor and the owner, but on public works projects in some states it is illegal as well.

Double Jeopardy

Many labor and materials payment bonds stipulate that the contractor must promptly make payment in full for all labor and materials used or required for the performance of a contract. This type of clause has been interpreted to mean that the general contractor can become liable for any unpaid bills of a subcontractor that has defaulted, even though the general contractor has paid the subcontractor for its just proportion of service or materials installed.

INSURANCE CHECKLIST

Part I—Liability Forms of Insurance

Worker's Compensation Insurance

Worker's compensation insurance protects employers from claims arising from various state laws relating to injuries sustained by employees in the course of employment. In some states, this coverage is underwritten by private companies; in others, there is a state fund for this purpose. State law in some areas may also require coverage for subcontractor employees if they are not otherwise covered. The following endorsements may be added to worker's compensation coverage:

1. *Longshoremen's and Harbor Workers' Compensation Act Liability* This endorsement should be added if there are operations on boats or docks.
2. *Extralegal or Additional Medical* If state provisions pertaining to medical expenses are not broad enough.
3. *Voluntary Compensation* Although the contractor may not be liable for accidents while its employees are not in the course of their employment, it may wish to cover them voluntarily, for example, when they are participating in company-sponsored athletics. State law may also permit this coverage for employees not otherwise protected under worker's compensation law.
4. *Universal (All States) Endorsement* If a contractor's employees travel away from their home state, they may become subject to the act or acts of other states. This coverage automatically covers contractor liability in such locations and furnishes coverage as well should liability arise under the Longshoremen's and Harbor Workers' Compensation Act.
5. *Special Maritime Endorsement* Covers work in connection with navigable waters.

Comprehensive General (Public) Liability Insurance

Comprehensive general liability insurance protects against legal liability to the public. There are many forms of liability insurance, but the one usually recommended is the Broad Form Comprehensive Liability Policy (automobiles included). All forms of liability insurance are combined in one contract. Physical damage may also be included on all owned automobiles. The following forms of liability insurance may or may not be included in the comprehensive form. If not, separate policies may be arranged:

1. *Premises–Operations* This coverage protects against the legal liability for bodily injury to persons other than employees and damage to the property of others that is not in the contractor's care, custody, or control. Exclusions should be checked carefully; for example, explosion, collapse, and underground damage are normally excluded from

coverage under the basic policy for most types of work. (These are usually designated exclusions "x," "c," and "u," respectively.) Although in some cases these exclusions cannot be removed, each project should be carefully examined for exposure to these hazards, and coverage secured when possible and necessary.

2 *Personal Injury* This protects against legal liability for claims arising from false arrest, libel, wrongful entry or eviction, and other related wrongs against a person. A check of the policy will reveal exclusions.

3 *Independent Contractors–Contractors Protective Liability* Coverage provides for the insured's legal liability that may arise from the acts or operations of a subcontractor or its employees, and damage to property of others if that property is not in the care, custody, or control of the contractor. Certificates of insurance should be secured from subcontractors and the scope of their insurance verified.

4 *Elevator* This is for permanent elevators such as those in a contractor's office. Temporary material hoists are covered under premises–operations.

5 *Contractual or Assumed Liability* Many construction contracts include a clause in which the contractor assumes the liability of someone else toward third parties. Such clauses can usually be recognized by the words "hold harmless" or "indemnify." It is recommended that the insurance policies of contractors and subcontractors provide "blanket contractual liability." Assumed liabilities are then covered automatically.

6 *Completed Operations (Products Liability)* This is an optional coverage that, subject to exclusions, protects against liability to persons or property of others that may arise after a project is completed; for example, from an accident due to faulty workmanship or materials. The actual replacement of faulty work cannot be covered.

7 *Umbrella Excess Liability* This insurance provides catastrophe coverage for claims in excess of the limits of liability afforded by other policies, and also for some hazards normally excluded in underlying liability policies. It may be subject to a large deductible feature ($10,000 to $25,000) and is a reasonably inexpensive way to protect a business from claims that could arise from a disastrous occurrence.

8 *Automobile* All liability from the existence or operation of any owned, hired, or nonowned vehicle may be included in this provision. This insurance should be on an "automatic" basis to provide

coverage for newly added equipment. A special endorsement may be needed if employees or their families use company cars. A "Use of Other Car" endorsement, naming each person so protected, would provide coverage for individuals using cars not owned by them or their employer.

9. *Automobile Medical Payments* This may be added to a policy. In some states, it is required.
10. *Automobile Physical Damage* This covers damage to the property of others and also may be endorsed to include comprehensive and collision coverage on owned vehicles. A high deductible of $500 or more can result in a considerable savings in premiums.

Part II—Property Forms of Insurance

Standard Builder's Risk Insurance

Standard Builder's Risk Insurance protects against physical damage to the insured property during the construction period resulting from any of the perils named in the policy. This coverage provides reimbursement based upon actual loss or damage rather than any legal liability that may be incurred. There are four principal methods used to establish amounts of coverage and to determine the premium:

1. *Completed Value* This method is based upon the assumption that the value of a project increases at a constant rate during the course of construction. Although the policy is written for the value of the completed project, the premium is based upon a reduced or average value. The dollar coverage provided is the actual value of completed work and stored materials at any given time. This form of Builder's Risk must be taken out at the start of construction. It is recommended that this method be used for the typical building project.
2. *Reporting Basis* The contractor must periodically notify the insurance carrier during construction of the increase in the value of a project. Coverage and premiums are based on the reported value. This method is advantageous when the completed value is low during most of the construction period, but increases very rapidly toward the end. However, failure to report an increase in value may result in a lack of proper coverage.

3 *Automatic Builder's Risk* This policy form gives a contractor temporary protection automatically, pending the issuance of a specific policy for each project.
4 *Ordinary Builder's Risk* This seldom used type of policy form is written for a fixed value. Coverage may be increased by endorsement at the request of the insured.

Within the framework of the methods just presented of writing a builder's policy, the following perils may be covered:

1 Fire and lightning.
2 *Extended Coverage* Covers windstorm, hail, riot, civil commotion, nonowned aircraft, smoke, and explosion (other than that from boilers, machinery, or piping).
3 *Vandalism and Malicious Mischief* Excludes pilferage, burglary, larceny, theft, and damage to glass (other than glass block).
4 *Additional Perils* The standard Builder's Risk policy may be endorsed to provide for specific additional perils. These may include collapse (not caused by design error, faulty materials, or workmanship), landslide, ground water, surface water (other than flood), sprinkler leakage, explosion or rupture of boilers, machinery or piping, breakage of glass, pilferage, and theft.

It may be possible to obtain endorsements or separate policies to cover the perils of flood and earthquake; however, such coverage can be difficult to obtain. For these perils, it is recommended that the contractor request the project owner to secure this coverage in conjunction with its permanent insurance for the completed structure.

Some of these additional perils can be covered by adding an "all risk" type endorsement, or a Multiple-Perils Builder's Risk policy (described next) may be preferred. These additional coverages generally require a deductible clause.

Multiple-Peril (All Risk) Builder's Risk Insurance

Multiple-Peril (All Risk) Builder's Risk Insurance is a nonstandard type of policy that provides similar but broader coverage than the standard Builder's Risk policy. Although the term "all risk" is widely used, all perils are *not* covered: It is a relative term denoting the broader-than-usual coverage. Generally, each insurance carrier writes its own form of multiple-peril insurance.

Rather than naming the perils insured, this type of policy insures against all risks of direct physical loss or damage to property from any external cause *except* those specifically excluded in the policy. Thus, coverage is determined by what is excluded, not included, and the policy must be checked closely to determine the coverage provided. Some forms will require a deductible clause for some of the perils covered.

This type of policy (written on a Completed-Value basis) is recommended over a Standard Builder's Risk policy *if* sufficient care is taken at the outset to make sure that all desirable coverage is included. Protection can be tailored to an individual contractor's needs.

Boiler, Machinery, and Power Plant Insurance

Boiler, machinery, and power plant insurance insures against damage caused by the explosion or rupture of steam boilers, turbines, piping, or associated machinery. It is written as a separate policy covering specific equipment and is usually a combined property insurance and public liability insurance policy. Temporary equipment of this nature should be insured by the contractor. When a project contains permanent machinery of this type, it is recommended that the owner purchase a policy as part of its permanent coverage *prior* to the start-up and testing of the equipment.

Floater Policies

Some portable machinery, tools, equipment, and materials are not covered by Builder's Risk policies. The term *floater* is used in the sense that the insurance follows the property wherever it might be located (possibly subject to territorial limits). In addition to special floater policy forms that may be available to meet special requirements, the following are examples of the more common types:

1. *Contractor's Equipment Floater* This insurance provides coverage on any or all equipment of the contractor (and optionally that of others in its possession), other than equipment moved regularly under its own power over highways (licensed or unlicensed). This coverage is usually written on a "named peril" basis, but may also be included under a multiple-peril plan.
2. *Transportation Floater* This provides coverage against damage to property (the contractor's or another's) while being transported. It may be obtained on a per trip, per project, or an annual basis.

3 *Installation Floater* This insures the contractor on a named peril or all-risk basis while he or she is installing or moving valuable equipment, materials, or machinery. Premiums can be reduced by using a deductible clause. (Steam Boiler and Machinery coverage only applies after machinery is installed and ready to use; hence, the need for this type of coverage.)

Part III—Miscellaneous Insurance

Fire and Explosion Legal Liability

Fire and Explosion Legal Liability insurance protects against legal liability for damage caused to property of others (buildings or contents) that is owned, rented, or under the contractor's control, and that is excluded from the regular liability policy. Contractual liability (hold harmless) is excluded from this coverage.

Water Legal Liability

Water Legal Liability insurance protects against legal liability arising from claims due to damage to others' property from the use or escape of water on property owned, rented, or under the contractor's control.

Multiperil Crime

Multiperil Crime coverage is available through several types of policies variously designated as "crime" policies. Included are the following:

1 Employee dishonesty
2 Loss inside the premises
3 Loss outside the premises
4 Money orders and counterfeit currency
5 Depositors' forgery
6 Paymaster robbery (broad form)
7 Credit card coverage

Policies for this protection may be available separately or in a single broad form policy.

Business Interruption

Business Interruption insurance is designed to provide income during a period of unforeseen business interruption.

Valuable Papers' Destruction

Valuable Papers' Destruction insurance protects against the loss, damage, or destruction of valuable papers, such as accounting records, deeds, mortgages, manuscripts, and abstracts, subject to stipulated limits. Coverage does not apply if loss is by misplacement, unexplained disappearance, wear and tear, deterioration, vermin, or war.

Accounts Receivable

Accounts Receivable coverage protects against loss caused by the inability to collect outstanding accounts through damage or destruction of records by fire or another insured hazard.

Part IV—General Comments

Right of Subrogation

Fire insurance policies as well as policies covering legal liability usually contain a clause commonly termed "right of subrogation." The doctrine of subrogation grants the insurer whatever rights the insured possessed at the time of an occurrence. Some policies include a definite statement that subrogation shall prevail, whereas others reserve the right of an insurance carrier to demand subrogation in the event of a loss. In either case, when such provisions appear in policies held by an owner, a contractor whose operations may have caused a loss to the owner may be sued by the insurance carrier for the amount of money it paid to the owner under the policy for settlement of the loss.

Several measures can be taken by contractors to avert this possibility. The contractor can obtain a letter from the owner specifically waiving subrogation rights, or a clause to that effect can be inserted in the contract documents. A joint waiver of subrogation can be executed, wherein both the owner and contractor waive these rights, and the contractor should obtain similar waivers from subcontractors. It is preferable that waivers cover both the construction and warranty period.

Another procedure is to have the contractor and subcontractors, as their respective interest may appear, designated as named insureds jointly with the owner in all policies, as subrogation is not possible between coinsureds.

Care, Custody, and Control

With reference to Property Damage Insurance policies, contractors should examine the implications of a "care, custody, and control" clause contained therein that might read as follows:

> It is further provided that insurance under this endorsement does not apply to damage to or destruction of property owned, leased, occupied, or used by, or in the "care, custody or control" of, the Assured or any employee of the Assured.

The exclusion applies especially to such damage caused to property that is immediately in the "care, custody, or control" of the contractor's employees. Should such property be more remotely under the possession of the contractor's employees, the danger of this exclusion becoming an issue is lessened.

This exclusion may be modified, and in special cases perhaps even removed for an additional premium.

APPENDIX B

Construction Industry Arbitration Rules— American Arbitration Association

Construction Industry Arbitration Rules

AMERICAN CONSULTING
ENGINEERS COUNCIL

AMERICAN INSTITUTE OF ARCHITECTS

AMERICAN SOCIETY OF
CIVIL ENGINEERS

AMERICAN SOCIETY OF
LANDSCAPE ARCHITECTS

AMERICAN SUBCONTRACTORS
ASSOCIATION

ASSOCIATED BUILDERS AND
CONTRACTORS, INC.

ASSOCIATED GENERAL CONTRACTORS

ASSOCIATED SPECIALTY
CONTRACTORS, INC.

CONSTRUCTION SPECIFICATIONS
INSTITUTE

NATIONAL ASSOCIATION OF
HOME BUILDERS

NATIONAL SOCIETY OF
PROFESSIONAL ENGINEERS

American
Arbitration
Association

*As amended
and in effect
February 1, 1984*

For the Submission of existing disputes:–

We, the undersigned parties, hereby agree to submit to arbitration under the Construction Industry Arbitration Rules of the American Arbitration Association the following controversy: (cite briefly). We further agree that the above controversy be submitted to (one) (three) Arbitrator(s) selected from the panels of Arbitrators of the American Arbitration Association. We further agree that we will faithfully observe this agreement and the Rules and that we will abide by and perform any award rendered by the Arbitrator(s) and that a judgment of the Court having jurisdiction may be entered upon the award.

Standard Arbitration Clause

Parties may refer to these Rules in their contracts. For this purpose, the following clause may be used:

Any controversy or claim arising out of or relating to this contract, or the breach thereof, shall be settled by arbitration in accordance with the Construction Industry Arbitration Rules of the American Arbitration Association, and judgment upon the award rendered by the Arbitrator(s) may be entered in any Court having jurisdiction thereof.

Introduction

Arbitration is the voluntary submission of a dispute to a disinterested person or persons for final determination. The American Arbitration Association (AAA) does not act as arbitrator. Its function is to administer arbitrations in accordance with the agreement of the parties and to maintain panels from which arbitrators may be chosen by the parties. Once designated, the arbitrator decides the issues and renders a final and binding award.

The American Arbitration Association shall establish and maintain as members of its National Panel of Arbitrators individuals competent to hear and determine disputes administered under the Construction Industry Arbitration Rules. The AAA shall consider for appointment to the Construction Industry Panel persons recommended by the National Construction Industry Arbitration Committee as qualified to serve by virtue of their experience in the construction field.

When an agreement to arbitrate is included in a construction contract, it may expedite peaceful settlement without the necessity of going to arbitration at all. Thus, the arbitration clause is a form of insurance against loss of good will.

Mediation

In appropriate cases, the parties may wish to submit their dispute to mediation. In mediation, the neutral mediator assists the parties in reaching a settlement but does not have the authority to make a binding decision or award. Mediation is administered by the AAA in accordance with the Construction Industry Mediation Rules. Copies of these Rules are available through all of the AAA's Regional Offices.

TABLE OF CONTENTS

1. Agreement of Parties 5
2. Name of Tribunal 5
3. Administrator 5
4. Delegation of Duties 5
5. National Panel of Arbitrators 5
6. Office of Tribunal 5
7. Initiation under an Arbitration Provision in a Contract 6
8. Change of Claim or Counterclaim 6
9. Initiation under a Submission 7
10. Pre-Hearing Conference and Preliminary Hearing . 7
11. Fixing of Locale 7
12. Qualifications of Arbitrator 7
13. Appointment from Panel 7
14. Direct Appointment by Parties 8
15. Appointment of Arbitrator by Party-Appointed Arbitrators 8
16. Nationality of Arbitrator in International Arbitration 9
17. Number of Arbitrators 9
18. Notice to Arbitrator of Appointment . 9
19. Disclosure and Challenge Procedure .. 9
20. Vacancies 10
21. Time and Place 10
22. Representation by Counsel 10
23. Stenographic Record 10
24. Interpreter 10
25. Attendance at Hearings 10
26. Adjournments 11
27. Oaths 11
28. Majority Decision 11
29. Order of Proceedings 11
30. Arbitration in the Absence of a Party or Counsel . 11
31. Evidence 12
32. Evidence by Affidavit and Filing of Documents ... 12
33. Inspection or Investigation 12
34. Conservation of Property 12
35. Closing of Hearings 13
36. Reopening of Hearings 13
37. Waiver of Oral Hearings 13
38. Waiver of Rules 13
39. Extensions of Time 13
40. Communication with Arbitrator and Serving of Notice 14
41. Time of Award 14
42. Form of Award 14
43. Scope of Award 14
44. Award upon Settlement 15
45. Delivery of Award to Parties 15
46. Release of Documents for Judicial Proceedings ... 15
47. Applications to Court and Exclusion of Liability 15
48. Administrative Fees 15
49. Fee When Oral Hearings are Waived . 16
50. Expenses 16
51. Arbitrator's Fee 16
52. Deposits 17
53. Interpretation and Application of Rules 17

EXPEDITED PROCEDURES

54. Notice by Telephone 17
55. Appointment and Qualifications of Arbitrators ... 17
56. Time and Place of Hearing 18
57. The Hearing 18
58. Time of Award 18
Administrative Fee Schedule 18

Construction Industry Arbitration Rules

1. Agreement of Parties
The parties shall be deemed to have made these Rules a part of their arbitration agreement whenever they have provided for arbitration under the Construction Industry Arbitration Rules. These Rules and any amendment thereof shall apply in the form obtaining at the time the arbitration is initiated.

2. Name of Tribunal
Any Tribunal constituted by the parties for the settlement of their dispute under these Rules shall be called the Construction Industry Arbitration Tribunal, hereinafter called the Tribunal.

3. Administrator
When parties agree to arbitrate under these Rules, or when they provide for arbitration by the American Arbitration Association, hereinafter called AAA, and an arbitration is initiated hereunder, they thereby constitute AAA the administrator of the arbitration. The authority and duties of the administrator are prescribed in the agreement of the parties and in these Rules.

4. Delegation of Duties
The duties of the AAA under these Rules may be carried out through Tribunal Administrators, or such other officers or committees as the AAA may direct.

5. National Panel of Arbitrators
In cooperation with the National Construction Industry Arbitration Committee, the AAA shall establish and maintain a National Panel of Construction Arbitrators, hereinafter called the Panel, and shall appoint an arbitrator or arbitrators therefrom as hereinafter provided. A neutral arbitrator selected by mutual choice of both parties or their appointees, or appointed by the AAA, is hereinafter called the arbitrator, whereas an arbitrator selected unilaterally by one party is hereinafter called the party-appointed arbitrator. The term arbitrator may hereinafter be used to refer to one arbitrator or to a Tribunal of multiple arbitrators.

6. Office of Tribunal
The general office of a Tribunal is the headquarters of the AAA, which may, however, assign the administration of an arbitration to any of its Regional Offices.

7. Initiation under an Arbitration Provision in a Contract
Arbitration under an arbitration provision in a contract shall be initiated in the following manner:

The initiating party shall, within the time specified by the contract, if any, file with the other party a notice of an intention to arbitrate (Demand), which notice shall contain a statement setting forth the nature of the dispute, the amount involved, and the remedy sought; and shall file three copies of said notice with any Regional Office of the AAA, together with three copies of the arbitration provisions of the contract and the appropriate filing fee as provided in Section 48 hereunder.

The AAA shall give notice of such filing to the other party. A party upon whom the demand for arbitration is made may file an answering statement in duplicate with the AAA within seven days after notice from the AAA, simultaneously sending a copy to the other party. If a monetary claim is made in the answer the appropriate administrative fee provided in the Fee Schedule shall be forwarded to the AAA with the answer. If no answer is filed within the stated time, it will be treated as a denial of the claim. Failure to file an answer shall not operate to delay the arbitration.

Unless the AAA in its discretion determines otherwise, the Expedited Procedures of Construction Arbitration shall be applied in any case where the total claim of any party does not exceed $15,000, exclusive of interest and arbitration costs. Parties may also agree to the Expedited Procedures in cases involving claims in excess of $15,000. The Expedited Procedures shall be applied as described in Sections 54 through 58 of these Rules.

8. Change of Claim or Counterclaim
After filing of the claim or counterclaim, if either party desires to make any new or different claim or counterclaim, same shall be made in writing and filed with the AAA, and a copy thereof shall be mailed to the other party who shall have a period of seven days from the date of such mailing within which to file an answer with the AAA. However, after the arbitrator is appointed no new or different claim or counterclaim may be submitted without the arbitrator's consent.

9. Initiation under a Submission

Parties to any existing dispute may commence an arbitration under these Rules by filing at any Regional Office two copies of a written agreement to arbitrate under these Rules (Submission), signed by the parties. It shall contain a statement of the matter in dispute, the amount of money involved, and the remedy sought, together with the appropriate filing fee as provided in the Fee Schedule.

10. Pre-Hearing Conference and Preliminary Hearing

At the request of the parties or at the discretion of the AAA, a pre-hearing conference with the administrator and the parties or their counsel will be scheduled in appropriate cases to arrange for an exchange of information and the stipulation of uncontested facts so as to expedite the arbitration proceedings.

In large and complex cases, unless the parties agree otherwise, the AAA may schedule a preliminary hearing with the parties and the arbitrator(s) to establish the extent of and schedule for the production of relevant documents and other information, the identification of any witnesses to be called, and a schedule for further hearings to resolve the dispute.

11. Fixing of Locale

The parties may mutually agree on the locale where the arbitration is to be held. If any party requests that the hearing be held in a specific locale and the other party files no objection thereto within seven days after notice of the request is mailed to such party, the locale shall be the one requested. If a party objects to the locale requested by the other party, the AAA shall have power to determine the locale and its decision shall be final and binding.

12. Qualifications of Arbitrator

Any arbitrator appointed pursuant to Section 13 or Section 15 shall be neutral, subject to disqualification for the reasons specified in Section 19. If the agreement of the parties names an arbitrator or specifies any other method of appointing an arbitrator, or if the parties specifically agree in writing, such arbitrator shall not be subject to disqualification for said reasons.

13. Appointment from Panel

If the parties have not appointed an arbitrator and have not provided any other method of appointment, the arbitrator shall be appointed in the following manner: Immediately after the filing of the Demand or Submission, the AAA shall submit simultaneously to each party to the dispute an identical list of names of persons chosen from the Panel. Each party to the dispute shall have seven days from the mailing date in which to cross off any names to which it objects, number the remaining names to indicate the order of preference, and return the list to the AAA. If a party does not return the list within the time specified, all persons named therein shall be deemed acceptable. From among the persons who have been approved on both lists, and in accordance with the designated order of mutual preference, the AAA shall invite the acceptance of an arbitrator to serve. If the parties fail to agree upon any of the persons named, or if acceptable arbitrators are unable to act, or if for any other reason the appointment cannot be made from the submitted lists, the AAA shall have the power to make the appointment from other members of the Panel without the submission of any additional lists.

14. Direct Appointment by Parties

If the agreement of the parties names an arbitrator or specifies a method of appointing an arbitrator, that designation or method shall be followed. The notice of appointment, with name and address of such arbitrator, shall be filed with the AAA by the appointing party. Upon the request of any such appointing party, the AAA shall submit a list of members of the Panel from which the party may make the appointment.

If the agreement specifies a period of time within which an arbitrator shall be appointed, and any party fails to make such appointment within that period, the AAA shall make the appointment.

If no period of time is specified in the agreement, the AAA shall notify the parties to make the appointment, and if within seven days after mailing of such notice such arbitrator has not been so appointed, the AAA shall make the appointment.

15. Appointment of Arbitrator by Party-Appointed Arbitrators

If the parties have appointed their party-appointed arbitrators or if either or both of them have been appointed as provided in Section 14, and have authorized such arbitrator to appoint an arbitrator within a specified time and no appointment is made within such time or any agreed extension thereof, the AAA shall appoint an arbitrator who shall act as Chairperson.

If no period of time is specified for appointment of the third arbitrator and the party-appointed arbitrators do not make the appointment within seven days from the date of the appointment of the last party-appointed arbitrator, the AAA shall appoint the arbitrator who shall act as Chairperson.

If the parties have agreed that their party-appointed arbitrators shall appoint the arbitrator from the Panel, the AAA shall furnish to the party-appointed arbitrators, in the manner prescribed in Section 13, a list selected from the Panel, and the appointment of the arbitrator shall be made as prescribed in such Section.

16. Nationality of Arbitrator in International Arbitration
If one of the parties is a national or resident of a country other than the United States, the arbitrator shall, upon the request of either party, be appointed from among the nationals of a country other than that of any of the parties.

17. Number of Arbitrators
If the arbitration agreement does not specify the number of arbitrators, the dispute shall be heard and determined by one arbitrator, unless the AAA, in its discretion, directs that a greater number of arbitrators be appointed.

18. Notice to Arbitrator of Appointment
Notice of the appointment of the arbitrator, whether mutually appointed by the parties or appointed by the AAA, shall be mailed to the arbitrator by the AAA, together with a copy of these Rules, and the signed acceptance of the arbitrator shall be filed prior to the opening of the first hearing.

19. Disclosure and Challenge Procedure
A person appointed as neutral arbitrator shall disclose to the AAA any circumstances likely to affect his or her impartiality, including any bias or any financial or personal interest in the result of the arbitration or any past or present relationship with the parties or their counsel. Upon receipt of such information from such arbitrator or other source, the AAA shall communicate such information to the parties and, if it deems it appropriate to do so, to the arbitrator and others. Thereafter, the AAA shall determine whether the arbitrator should be disqualified and shall inform the parties of its decision, which shall be conclusive.

20. Vacancies
If any arbitrator should resign, die, withdraw, refuse, be disqualified or be unable to perform the duties of office, the AAA shall, on proof satisfactory to it, declare the office vacant. Vacancies shall be filled in accordance with the applicable provision of these Rules. In the event of a vacancy in a panel of arbitrators, the remaining arbitrator or arbitrators may continue with the hearing and determination of the controversy, unless the parties agree otherwise.

21. Time and Place
The arbitrator shall fix the time and place for each hearing. The AAA shall mail to each party notice thereof at least five days in advance, unless the parties by mutual agreement waive such notice or modify the terms thereof.

22. Representation by Counsel
Any party may be represented by counsel. A party intending to be so represented shall notify the other party and the AAA of the name and address of counsel at least three days prior to the date set for the hearing at which counsel is first to appear. When an arbitration is initiated by counsel, or where an attorney replies for the other party, such notice is deemed to have been given.

23. Stenographic Record
Any party wishing a stenographic record shall make such arrangements directly with the stenographer and shall notify the other parties of such arrangements in advance of the hearing. The requesting party or parties shall pay the cost of such record.

24. Interpreter
Any party wishing an interpreter shall make all arrangements directly with an interpreter and shall assume the costs of such service.

25. Attendance at Hearings
Persons having a direct interest in the arbitration are entitled to attend hearings. The arbitrator shall otherwise have the power to require the retirement of any witness or witnesses during the testimony of other witnesses. It shall be discretionary with the arbitrator to determine the propriety of the attendance of any other persons.

26. Adjournments
The arbitrator may adjourn the hearing, and must take such adjournment when all of the parties agree thereto.

27. Oaths
Before proceeding with the first hearing or with the examination of the file, each arbitrator may take an oath of office, and if required by law, shall do so. The arbitrator may require witnesses to testify under oath administered by any duly qualified person or, if required by law or demanded by either party, shall do so.

28. Majority Decision
Whenever there is more than one arbitrator, all decisions of the arbitrators must be by at least a majority. The award must also be made by at least a majority unless the concurrence of all is expressly required by the arbitration agreement or by law.

29. Order of Proceedings
A hearing shall be opened by the filing of the oath of the arbitrator, where required, and by the recording of the place, time, and date of the hearing, the presence of the arbitrator and parties, and counsel, if any, and by the receipt by the arbitrator of the statement of the claim and answer, if any.

The arbitrator may, at the beginning of the hearing, ask for statements clarifying the issues involved. In some cases, part or all of the above will have been accomplished at the preliminary hearing conducted by the arbitrator(s) pursuant to Section 10.

The complaining party shall then present its claims, proofs and witnesses, who shall submit to questions or other examination. The defending party shall then present its defenses, proofs and witnesses, who shall submit to questions or other examination. The arbitrator may vary this procedure but shall afford full and equal opportunity to the parties for the presentation of any material or relevant proofs.

Exhibits, when offered by either party, may be received in evidence by the arbitrator.

The names and addresses of all witnesses and exhibits in order received shall be made a part of the record.

30. Arbitration in the Absence of a Party or Counsel
Unless the law provides to the contrary, the arbitration may proceed in the absence of any party or counsel, who, after due notice, fails to be present or fails to obtain an adjournment. An award shall not be made solely on the default of a party. The arbitrator shall require the party who is present to submit such evidence as is deemed necessary for the making of an award.

31. Evidence
The parties may offer such evidence as is pertinent and material to the controversy and shall produce such additional evidence as the arbitrator may deem necessary to an understanding and determination of the controversy. An arbitrator authorized by law to subpoena witnesses or documents may do so upon the request of any party, or independently.

The arbitrator shall be the judge of the relevance and the materiality of the evidence offered, and conformity to legal rules of evidence shall not be necessary. All evidence shall be taken in the presence of all of the arbitrators and all of the parties, except where any of the parties is absent in default or has waived the right to be present.

32. Evidence by Affidavit and Filing of Documents
The arbitrator may receive and consider the evidence of witnesses by affidavit, giving it such weight as seems appropriate after consideration of any objections made to its admission.

All documents not filed with the arbitrator at the hearing, but arranged for at the hearing or subsequently by agreement of the parties, shall be filed with the AAA for transmission to the arbitrator. All parties shall be afforded opportunity to examine such documents.

33. Inspection or Investigation
An arbitrator finding it necessary to make an inspection or investigation in connection with the arbitration shall direct the AAA to so advise the parties. The arbitrator shall set the time and the AAA shall notify the parties thereof. Any party who so desires may be present at such inspection or investigation. In the event that one or both parties are not present at the inspection or investigation, the arbitrator shall make a verbal or written report to the parties and afford them an opportunity to comment.

34. Conservation of Property
The arbitrator may issue such orders as may be

deemed necessary to safeguard the property which is the subject matter of the arbitration without prejudice to the rights of the parties or to the final determination of the dispute.

35. Closing of Hearings
The arbitrator shall specifically inquire of the parties whether they have any further proofs to offer or witnesses to be heard. Upon receiving negative replies, the arbitrator shall declare the hearings closed and a minute thereof shall be recorded. If briefs are to be filed, the hearings shall be declared closed as of the final date set by the arbitrator for the receipt of briefs. If documents are to be filed as provided for in Section 32 and the date set for their receipt is later than that set for the receipt of briefs, the later date shall be the date of closing the hearing. The time limit within which the arbitrator is required to make an award shall commence to run, in the absence of other agreements by the parties, upon the closing of the hearings.

36. Reopening of Hearings
The hearings may be reopened by the arbitrator at will, or upon application of a party at any time before the award is made. If the reopening of the hearing would prevent the making of the award within the specific time agreed upon by the parties in the contract out of which the controversy has arisen, the matter may not be reopened, unless the parties agree upon the extension of such time limit. When no specific date is fixed in the contract, the arbitrator may reopen the hearings, and the arbitrator shall have thirty days from the closing of the reopened hearings within which to make an award.

37. Waiver of Oral Hearings
The parties may provide, by written agreement, for the waiver of oral hearings. If the parties are unable to agree as to the procedure, the AAA shall specify a fair and equitable procedure.

38. Waiver of Rules
Any party who proceeds with the arbitration after knowledge that any provision or requirement of these Rules has not been complied with and who fails to state an objection thereto in writing, shall be deemed to have waived the right to object.

39. Extensions of Time
The parties may modify any period of time by mutual agreement. The AAA for good cause may extend any period of time established by these Rules, except the time for making the award. The AAA shall notify the parties of any such extension of time and its reason therefor.

40. Communication with Arbitrator and Serving of Notices
There shall be no communication between the parties and an arbitrator other than at oral hearings. Any other oral or written communications from the parties to the arbitrator shall be directed to the AAA for transmittal to the arbitrator.

Each party to an agreement which provides for arbitration under these Rules shall be deemed to have consented that any papers, notices or process necessary or proper for the initiation or continuation of an arbitration under these Rules and for any court action in connection therewith or for the entry of judgment on any award made thereunder may be served upon such party by mail addressed to such party or its attorney at the last known address or by personal service, within or without the state wherein the arbitration is to be held (whether such party be within or without the United States of America), provided that reasonable opportunity to be heard with regard thereto has been granted such party.

41. Time of Award
The award shall be made promptly by the arbitrator and, unless otherwise agreed by the parties, or specified by law, not later than thirty days from the date of closing the hearings, or if oral hearings have been waived, from the date of transmitting the final statements and proofs to the arbitrator.

42. Form of Award
The award shall be in writing and shall be signed either by the sole arbitrator or by at least a majority if there be more than one. It shall be executed in the manner required by law.

43. Scope of Award
The arbitrator may grant any remedy or relief which is just and equitable and within the terms of the agreement of the parties. The arbitrator, in the award, shall assess arbitration fees and expenses as provided in Sections 48 and 50 equally or in favor of any party and, in the event any administrative fees or expenses are due the AAA, in favor of the AAA.

44. Award upon Settlement
If the parties settle their dispute during the course of the arbitration, the arbitrator, upon their request, may set forth the terms of the agreed settlement in an award.

45. Delivery of Award to Parties
Parties shall accept as legal delivery of the award the placing of the award or a true copy thereof in the mail by the AAA, addressed to such party at its last known address or to its attorney, or personal service of the award, or the filing of the award in any manner which may be prescribed by law.

46. Release of Documents for Judicial Proceedings
The AAA shall, upon the written request of a party, furnish to such party, at its expense, certified facsimiles of any papers in the AAA's possession that may be required in judicial proceedings relating to the arbitration.

47. Applications to Court and Exclusion of Liability
(a) No judicial proceedings by a party relating to the subject matter of the arbitration shall be deemed a waiver of the party's right to arbitrate.

(b) Neither the AAA nor any arbitrator in a proceeding under these Rules is a necessary party in judicial proceedings relating to the arbitration.

(c) Parties to these Rules shall be deemed to have consented that judgment upon the award rendered by the arbitrator(s) may be entered in any Federal or State Court having jurisdiction thereof.

(d) Neither the AAA nor any arbitrator shall be liable to any party for any act or omission in connection with any arbitration conducted under these Rules.

48. Administrative Fees
As a not-for-profit organization, the AAA shall prescribe an administrative fee schedule and a refund schedule to compensate it for the cost of providing administrative services. The schedule in effect at the time of filing or the time of refund shall be applicable.

The administrative fees shall be advanced by the initiating party or parties in accordance with the administrative fee schedule, subject to final apportionment by the arbitrator in the award.

When a matter is withdrawn or settled, the refund shall be made in accordance with the refund schedule.

The AAA, in the event of extreme hardship on the part of any party, may defer or reduce the administrative fee.

49. Fee When Oral Hearings are Waived
Where all oral hearings are waived under Section 37, the Administrative Fee Schedule shall apply.

50. Expenses
The expenses of witnesses for either side shall be paid by the party producing such witnesses.

The cost of the stenographic record, if any is made, and all transcripts thereof, shall be prorated equally between the parties ordering copies, unless they shall otherwise agree, and shall be paid for by the responsible parties directly to the reporting agency.

All other expenses of the arbitration, including required traveling and other expenses of the arbitrator and of AAA representatives, and the expenses of any witness or the cost of any proofs produced at the direct request of the arbitrator, shall be borne equally by the parties, unless they agree otherwise, or unless the arbitrator in the award assesses such expenses or any part thereof against any specified party or parties.

51. Arbitrator's Fee
Unless the parties agree to terms of compensation, members of the National Panel of Construction Arbitrators will serve without compensation for the first two days of service.

Thereafter, compensation shall be based upon the amount of service involved and the number of hearings. An appropriate daily rate and other arrangements will be discussed by the administrator with the parties and the arbitrator(s). If the parties fail to agree to the terms of compensation, an appropriate rate shall be established by the AAA, and communicated in writing to the parties.

Any arrangement for the compensation of an arbitrator shall be made through the AAA and not directly by the arbitrator with the parties. The terms of compensation of neutral arbitrators on a Tribunal shall be identical.

52. Deposits
The AAA may require the parties to deposit in advance such sums of money as it deems necessary to defray the expense of the arbitration, including the arbitrator's fee, if any, and shall render an accounting to the parties and return any unexpended balance.

53. Interpretation and Application of Rules
The arbitrator shall interpret and apply these Rules insofar as they relate to the arbitrator's powers and duties. When there is more than one arbitrator and a difference arises among them concerning the meaning or application of any such Rules, it shall be decided by a majority vote. If that is unobtainable, either an arbitrator or a party may refer the question to the AAA for final decision. All other Rules shall be interpreted and applied by the AAA.

EXPEDITED PROCEDURES
54. Notice by Telephone
The parties shall accept all notices from the AAA by telephone. Such notices by the AAA shall subsequently be confirmed in writing to the parties. Notwithstanding the failure to confirm in writing any notice or objection hereunder, the proceeding shall nonetheless be valid if notice has, in fact, been given by telephone.

55. Appointment and Qualifications of Arbitrators
The AAA shall submit simultaneously to each party to the dispute an identical list of five members of the Construction Arbitration Panel of Arbitrators, from which one arbitrator shall be appointed. Each party shall have the right to strike two names from the list on a peremptory basis. The list is returnable to the AAA within ten days from the date of mailing. If for any reason the appointment cannot be made from the list, the AAA shall have the authority to make the appointment from among other members of the Panel without the submission of additional lists. Such appointment shall be subject to disqualification for the reasons specified in Section 19. The parties shall be given notice by telephone by the AAA of the appointment of the arbitrator. The parties shall notify the AAA, by telephone, within seven days of any objections to the arbitrator appointed. Any objection by a party to such arbitrator shall be confirmed in writing to the AAA with a copy to the other party(ies).

56. Time and Place of Hearing
The arbitrator shall fix the date, time, and place of the hearing. The AAA will notify the parties by telephone, seven days in advance of the hearing date. Formal Notice of Hearing will be sent by the AAA to the parties.

57. The Hearing
Generally, the hearing and presentations of the parties shall be completed within one day. The arbitrator, for good cause shown, may schedule an additional hearing to be held within five days.

58. Time of Award
Unless otherwise agreed to by the parties, the award shall be rendered not later than five business days from the date of the closing of the hearing.

ADMINISTRATIVE FEE SCHEDULE

A filing fee of $200 will be paid at the time the case is initiated.

The balance of the administrative fee of the AAA is based upon the amount of each claim and counterclaim as disclosed when the claim and counterclaim are filed, and is due and payable prior to the notice of appointment of the neutral arbitrator.

In those claims and counterclaims which are not for a monetary amount, an appropriate administrative fee will be determined by the AAA, payable prior to such notice of appointment.

Amount of Claim or Counterclaim	Fee for Claim or Counterclaim
$1 to $20,000	3% (minimum $200)
$20,000 to $40,000	$ 600, plus 2% of excess over $20,000
$40,000 to $80,000	$1,000, plus 1% of excess over $40,000
$80,000 to $160,000	$1,400, plus ½% of excess over $80,000
$160,000 to $5,000,000	$1,800, plus ¼% of excess over $160,000

Where the claim or counterclaim exceeds $5 million, an appropriate fee will be determined by the AAA. If there are more than two parties represented in the arbitration, an additional 10% of the administrative fee will be due for each additional represented party.

When no amount can be stated at the time of filing, the administrative fee is $500, subject to adjustment in accordance with the schedule as soon as an amount can be disclosed.

OTHER SERVICE CHARGES

$50 payable by each party for each second and subsequent hearing which is either clerked by the AAA or held in a hearing room provided by the AAA.

POSTPONEMENT FEES

Sole-Arbitrator Cases:
$50 payable by a party first causing an adjournment of any scheduled hearing.

$100 payable by a party causing a second or subsequent adjournment of any scheduled hearing.

Three-Arbitrator Cases:
$75 payable by a party first causing an adjournment of any scheduled hearing.

$150 payable by a party causing a second or subsequent adjournment of any scheduled hearing.

REFUND SCHEDULE

If the AAA is notified that a case has been settled or withdrawn before a list of Arbitrators has been sent out, all the fee in excess of $200 will be refunded.

If the AAA is notified that a case has been settled or withdrawn before the original due date for the return of the first list, two-thirds of the fee in excess of $200 will be refunded.

If the AAA is notified that a case is settled or withdrawn during or following a pre-hearing conference or at least 48 hours before the date and time set for the first hearing, one-third of the fee in excess of $200 will be refunded.

REGIONAL DIRECTORS

ATLANTA (30361) • INDIA JOHNSON •
1197 Peachtree Street, N.E. • (404) 872-3022

BOSTON (02114) • RICHARD M. REILLY •
60 Staniford Street • (617) 367-6800

CHARLOTTE (28226) • MARK SHOLANDER •
7301 Carmel Executive Park • (704) 541-1367

CHICAGO (60606) • LaVERNE ROLLE •
205 West Wacker Drive • (312) 346-2282

CINCINNATI (45202) • PHILIP S. THOMPSON •
2308 Carew Tower • (513) 241-8434

CLEVELAND (44115) • EARLE C. BROWN •
1127 Euclid Avenue • (216) 241-4741

DALLAS (75201) • HELMUT O. WOLFF •
1607 Main Street • (214) 748-4979

DENVER (80203) • MARK APPEL •
1775 Sherman Street • (303) 831-0823

DETROIT (48226) • MARY A. BEDIKIAN •
615 Griswold Street • (313) 964-2525

GARDEN CITY, N.Y. (11530) • MARK A. RESNICK •
585 Stewart Avenue • (516) 222-1660

HARTFORD (06106) • KAREN M. JALKUT •
2 Hartford Square West • (203) 278-5000

KANSAS CITY, MO. (64106) • NEIL MOLDENHAUER •
1101 Walnut Street • (816) 221-6401

LOS ANGELES (90020) • JERROLD L. MURASE •
443 Shatto Place • (213) 383-6516

MIAMI (33129) • RENE GRAFALS •
2250 S.W. 3rd Avenue • (305) 854-1616

MINNEAPOLIS (55402) • JAMES R. DEYE •
510 Foshay Tower • (612) 332-6545

NEW JERSEY (SOMERSET 08873) • RICHARD NAIMARK •
1 Executive Drive • (201) 560-9560

NEW YORK (10020) • GEORGE H. FRIEDMAN •
140 West 51st Street • (212) 484-4000

PHILADELPHIA (19102) • ARTHUR R. MEHR •
230 South Broad Street • (215) 732-5260

PHOENIX (85012) • DEBORAH A. KRELL •
77 East Columbus • (602) 234-0950

PITTSBURGH (15222) • JOHN F. SCHANO •
221 Gateway Four • (412) 261-3617

SAN DIEGO (92101) • DENNIS SHARP •
530 Broadway • (619) 239-3051

SAN FRANCISCO (94108) • CHARLES A. COOPER •
445 Bush Street • (415) 981-3901

SEATTLE (98104) • NEAL M. BLACKER •
811 First Avenue • (206) 622-6435

SYRACUSE (13202) • DEBORAH A. BROWN •
720 State Tower Building • (315) 472-5483

WASHINGTON, D.C. (20036) • GARYLEE COX •
1730 Rhode Island Avenue, N.W. • (202) 296-8510

WHITE PLAINS, N.Y. (10601) • MARION J. ZINMAN •
34 South Broadway • (914) 946 1119

AMERICAN ARBITRATION ASSOCIATION
NEW YORK (10020-1203) • 140 West 51st Street
(212) 484-4000

Sample Public Works Contract Documents (Contract Documents for Construction of Federally Assisted Water and Sewer Projects)

APPENDIX · C ·

CONTRACT DOCUMENTS
for
CONSTRUCTION OF FEDERALLY ASSISTED WATER AND SEWER PROJECTS

List of Documents

1. Advertisement for Bids
2. Information for Bidders
3. Bid
4. Bid Bond
5. Agreement
6. Payment Bond
7. Performance Bond
8. Notice of Award
9. Notice to Proceed
10. Change Order
11. General Conditions

PREFACE

These Contract Documents are acceptable for use by borrowers and grantees in Federally assisted projects funded by the below listed Federal agencies.

Local or state legislation may prohibit the use of some sections. The substitution or revision of individual sections, therefore, may be deemed appropriate.

Jointly prepared and endorsed by:
  Economic Development Administration, Department of Commerce
  Environmental Protection Agency
  Farmers Home Administration, Department of Agriculture
  Department of Housing and Urban Development
  American Consulting Engineers Council
  American Public Works Association
  American Society of Civil Engineers
  Associated General Contractors of America
  National Society of Professional Engineers
  National Utility Contractors Association

ADVERTISEMENT FOR BIDS

_____
Owner

_____
Address

 Separate sealed BIDS for the construction of (briefly describe nature, scope, and major elements of the work)_____

_____

_____

will be received by _____

at the office of _____

until _____, (Standard Time – Daylight Savings Time) _____,

19_____, and then at said office publicly opened and read aloud.

 The CONTRACT DOCUMENTS may be examined at the following locations:

_____

_____

_____

 Copies of the CONTRACT DOCUMENTS may be obtained at the office of _____

_____located at _____

upon payment of $_____ for each set.

 Any BIDDER, upon returning the CONTRACT DOCUMENTS promptly and in good condition, will be refunded his payment, and any non-bidder upon so returning the CONTRACT DOCUMENTS will be refunded $_____.

_____ _____
     Date

CONTRACT DOCUMENTS FOR CONSTRUCTION OF
FEDERALLY ASSISTED WATER AND SEWER PROJECTS

INFORMATION FOR BIDDERS

BIDS will be received by _____ (herein called the "OWNER"), at_____ until_____, 19_____, and then at said office publicly opened and read aloud.

Each BID must be submitted in a sealed envelope, addressed to_____ _____ at _____. Each sealed envelope containing a BID must be plainly marked on the outside as BID for _____ and the envelope should bear on the outside the name of the BIDDER, his address, his license number if applicable and the name of the project for which the BID is submitted. If forwarded by mail, the sealed envelope containing the BID must be enclosed in another envelope addressed to the OWNER at_____ _____.

All BIDS must be made on the required BID form. All blank spaces for BID prices must be filled in, in ink or typewritten, and the BID form must be fully completed and executed when submitted. Only one copy of the BID form is required.

The OWNER may waive any informalities or minor defects or reject any and all BIDS. Any BID may be withdrawn prior to the above scheduled time for the opening of BIDS or authorized postponement thereof. Any BID received after the time and date specified shall not be considered. No BIDDER may withdraw a BID within 60 days after the actual date of the opening thereof. Should there be reasons why the contract cannot be awarded within the specified period, the time may be extended by mutual agreement between the OWNER and the BIDDER.

BIDDERS must satisfy themselves of the accuracy of the estimated quantities in the BID Schedule by examination of the site and a review of the drawings and specifications including ADDENDA. After BIDS have been submitted, the BIDDER shall not assert that there was a misunderstanding concerning the quantities of WORK or of the nature of the WORK to be done.

The OWNER shall provide to BIDDERS prior to BIDDING, all information which is pertinent to, and delineates and describes, the land owned and rights-of-way acquired or to be acquired.

The CONTRACT DOCUMENTS contain the provisions required for the construction of the PROJECT. Information obtained from an officer, agent, or employee of the OWNER or any other person shall not affect the risks or obligations assumed by the CONTRACTOR or relieve him from fulfilling any of the conditions of the contract.

Each BID must be accompanied by a BID bond payable to the OWNER for five percent of the total amount of the BID. As soon as the BID prices have been compared, the OWNER will return the BONDS of all except the three lowest responsible BIDDERS. When the Agreement is executed the bonds of the two remaining unsuccessful BIDDERS will be returned. The BID BOND of the successful BIDDER will be retained until the payment BOND and performance BOND have been executed and approved, after which it will be returned. A certified check may be used in lieu of a BID BOND.

CONTRACT DOCUMENTS FOR CONSTRUCTION OF
FEDERALLY ASSISTED WATER AND SEWER PROJECTS

Document No. 2
Information for Bidders: page 1 of 2

A performance BOND and a payment BOND, each in the amount of 100 percent of the CONTRACT PRICE, with a corporate surety approved by the OWNER, will be required for the faithful performance of the contract.

Attorneys-in-fact who sign BID BONDS or payment BONDS and performance BONDS must file with each BOND a certified and effective dated copy of their power of attorney.

The party to whom the contract is awarded will be required to execute the Agreement and obtain the performance BOND and payment BOND within ten (10) calendar days from the date when NOTICE OF AWARD is delivered to the BIDDER. The NOTICE OF AWARD shall be accompanied by the necessary Agreement and BOND forms. In case of failure of the BIDDER to execute the Agreement, the OWNER may at his option consider the BIDDER in default, in which case the BID BOND accompanying the proposal shall become the property of the OWNER.

The OWNER within ten (10) days of receipt of acceptable performance BOND, payment BOND and Agreement signed by the party to whom the Agreement was awarded shall sign the Agreement and return to such party an executed duplicate of the Agreement. Should the OWNER not execute the Agreement within such period, the BIDDER may by WRITTEN NOTICE withdraw his signed Agreement. Such notice of withdrawal shall be effective upon receipt of the notice by the OWNER.

The NOTICE TO PROCEED shall be issued within ten (10) days of the execution of the Agreement by the OWNER. Should there be reasons why the NOTICE TO PROCEED cannot be issued within such period, the time may be extended by mutual agreement between the OWNER and CONTRACTOR. If the NOTICE TO PROCEED has not been issued within the ten (10) day period or within the period mutually agreed upon, the CONTRACTOR may terminate the Agreement without further liability on the part of either party.

The OWNER may make such investigations as he deems necessary to determine the ability of the BIDDER to perform the WORK, and the BIDDER shall furnish to the OWNER all such information and data for this purpose as the OWNER may request. The OWNER reserves the right to reject any BID if the evidence submitted by, or investigation of, such BIDDER fails to satisfy the OWNER that such BIDDER is properly qualified to carry out the obligations of the Agreement and to complete the WORK contemplated therein.

A conditional or qualified BID will not be accepted.

Award will be made to the lowest responsible BIDDER.

All applicable laws, ordinances, and the rules and regulations of all authorities having jurisdiction over construction of the PROJECT shall apply to the contract throughout.

Each BIDDER is responsible for inspecting the site and for reading and being thoroughly familiar with the CONTRACT DOCUMENTS. The failure or omission of any BIDDER to do any of the foregoing shall in no way relieve any BIDDER from any obligation in respect to his BID.

Further, the BIDDER agrees to abide by the requirements under Executive Order No. 11246, as amended, including specifically the provisions of the equal opportunity clause set forth in the SUPPLEMENTAL GENERAL CONDITIONS.

The low BIDDER shall supply the names and addresses of major material SUPPLIERS and SUBCONTRACTORS when requested to do so by the OWNER.

Inspection trips for prospective BIDDERS will leave from the office of the _____ at _____

The ENGINEER is _____. His address is _____

Document No. 2
Information for Bidders: page 2 of 2

BID

Proposal of _____ (hereinafter called "BIDDER"), organized and existing under the laws of the State of _____ doing business as _____*.

To the _____

_____(hereinafter called "OWNER").

In compliance with your Advertisement for Bids, BIDDER hereby proposes to perform all WORK for the construction of _____

_____

in strict accordance with the CONTRACT DOCUMENTS, within the time set forth therein, and at the prices stated below.

By submission of this BID, each BIDDER certifies, and in the case of a joint BID each party thereto certifies as to his own organization, that this BID has been arrived at independently, without consultation, communication, or agreement as to any matter relating to this BID with any other BIDDER or with any competitor.

BIDDER hereby agrees to commence WORK under this contract on or before a date to be specified in the NOTICE TO PROCEED and to fully complete the PROJECT within _____ consecutive calendar days thereafter. BIDDER further agrees to pay as liquidated damages, the sum of $_____ for each consecutive calendar day thereafter as provided in Section 15 of the General Conditions.

BIDDER acknowledges receipt of the following ADDENDUM:

_____
_____
_____

*Insert "a corporation", "a partnership", or "an individual" as applicable.

CONTRACT DOCUMENTS FOR CONSTRUCTION OF
FEDERALLY ASSISTED WATER AND SEWER PROJECTS

Document No. 3
Bid: page 1 of 3

BIDDER agrees to perform all the work described in the CONTRACT DOCUMENTS for the following unit prices or lump sum:

BID SCHEDULE

NOTE: BIDS shall include sales tax and all other applicable taxes and fees.

NO.	ITEM	UNIT	UNIT PRICE	AMOUNT	TOTAL PRICE

NO.	ITEM	UNIT	UNIT PRICE	AMOUNT	TOTAL PRICE

TOTAL OF BID .... $_____
LUMP SUM PRICE (if applicable) .... $_____

    Respectfully submitted:

_____ _____
     Signature   Address

_____ _____
      Title   Date

_____
 License Number (if applicable)

(SEAL — if BID is by a corporation)

Attest _____

Document No. 3
Bid: page 3 of 3

BID BOND

KNOW ALL MEN BY THESE PRESENTS, that we, the undersigned, _____
_____ as Principal, and
_____ as Surety, are hereby
held and firmly bound unto _____ as OWNER
in the penal sum of _____
for the payment of which, well and truly to be made, we hereby jointly and severally bind ourselves, successors and assigns.

Signed, this _____ day of _____, 19_____.

The Condition of the above obligation is such that whereas the Principal has submitted to _____ a certain BID, attached hereto and hereby made a part hereof to enter into a contract in writing, for the

_____
_____
_____

NOW, THEREFORE,

    (a) If said BID shall be rejected, or

    (b) If said BID shall be accepted and the Principal shall execute and deliver a contract in the Form of Contract attached hereto (properly completed in accordance with said BID) and shall furnish a BOND for his faithful performance of said contract, and for the payment of all persons performing labor or furnishing materials in connection therewith, and shall in all other respects perform the agreement created by the acceptance of said BID,

then this obligation shall be void, otherwise the same shall remain in force and effect; it being expressly understood and agreed that the liability of the Surety for any and all claims hereunder shall, in no event, exceed the penal amount of this obligation as herein stated.

The Surety, for value received, hereby stipulates and agrees that the obligations of said Surety and its BOND shall be in no way impaired or affected by any extension of the time within which the OWNER may accept such BID; and said Surety does herby waive notice of any such extension.

IN WITNESS WHEREOF, the Principal and the Surety have hereunto set their hands and seals, and such of them as are corporations have caused their corporate seals to be hereto affixed and these presents to be signed by their proper officers, the day and year first set forth above.

_____ (L.S.)
　　　　　　Principal

_____
　　　　　　Surety

By: _____

IMPORTANT — Surety companies executing BONDS must appear on the Treasury Department's most current list (Circular 570 as amended) and be authorized to transact business in the state where the project is located.

Document No. 4
Bid Bond: Page 2 of 2

AGREEMENT

THIS AGREEMENT, made this _____ day of _____, 19_____, by and between _____, hereinafter called "OWNER"
(Name of Owner), (an Individual)

and _____ doing business as (an individual,) or (a partnership,) or (a corporation) hereinafter called "CONTRACTOR".

WITNESSETH: That for and in consideration of the payments and agreements hereinafter mentioned:

    1. The CONTRACTOR will commence and complete the construction of

_____

    2. The CONTRACTOR will furnish all of the material, supplies, tools, equipment, labor and other services necessary for the construction and completion of the PROJECT described herein.

    3. The CONTRACTOR will commence the work required by the CONTRACT DOCUMENTS within _____ calendar days after the date of the NOTICE TO PROCEED and will complete the same within _____ calendar days unless the period for completion is extended otherwise by the CONTRACT DOCUMENTS.

    4. The CONTRACTOR agrees to perform all of the WORK described in the CONTRACT DOCUMENTS and comply with the terms therein for the sum of $ _____, or as shown in the BID schedule.

    5. The term "CONTRACT DOCUMENTS" means and includes the following:

    (A) Advertisement For BIDS

    (B) Information For BIDDERS

    (C) BID

    (D) BID BOND

    (E) Agreement

(F) General Conditions

(G) SUPPLEMENTAL GENERAL CONDITIONS

(H) Payment BOND

(I) Performance BOND

(J) NOTICE OF AWARD

(K) NOTICE TO PROCEED

(L) CHANGE ORDER

(M) DRAWINGS prepared by _____
numbered _____ through _____, and dated _____,
19_____

(N) SPECIFICATIONS prepared or issued by _____,
dated _____, 19_____

(O) ADDENDA:

No. _____, dated _____, 19_____

No. _____, dated _____, 19_____

No. _____, dated _____, 19_____

No. _____, dated _____, 19_____

No. _____, dated _____, 19_____

No. _____, dated _____, 19_____

6. The OWNER will pay to the CONTRACTOR in the manner and at such times as set forth in the General Conditions such amounts as required by the CONTRACT DOCUMENTS.

7. This Agreement shall be binding upon all parties hereto and their respective heirs, executors, administrators, successors, and assigns.

IN WITNESS WHEREOF, the parties hereto have executed, or caused to be executed by their duly authorized officials, this Agreement in (_____) each of
(Number of Copies)
which shall be deemed an original on the date first above written.

Document No. 5
Agreement: Page 2 of 3

OWNER:
_____

BY _____

Name _____
                (Please Type)

Title _____

(SEAL)
ATTEST:
_____

Name _____
          (Please Type)

Title _____

CONTRACTOR:
_____

BY _____

Name _____
             (Please Type)

Address _____
_____

(SEAL)
ATTEST:
_____

Name _____
          (Please Type)
_____

PAYMENT BOND

KNOW ALL MEN BY THESE PRESENTS: that

_____
(Name of Contractor)

_____
(Address of Contractor)

a _____, hereinafter called Principal,
   (Corporation, Partnership or Individual)

and _____
(Name of Surety)

_____
(Address of Surety)

hereinafter called Surety, are held and firmly bound unto _____

_____
(Name of Owner)

_____
(Address of Owner)

hereinafter called OWNER, in the penal sum of _____ Dollars, $(_____)
in lawful money of the United States, for the payment of which sum well and truly to be made, we bind ourselves, successors, and assigns, jointly and severally, firmly by these presents.

THE CONDITION OF THIS OBLIGATION is such that whereas, the Principal entered into a certain contract with the OWNER, dated the _____ day of _____ 19____, a copy of which is hereto attached and made a part hereof for the construction of:

_____

_____

_____

NOW, THEREFORE, if the Principal shall promptly make payment to all persons, firms, SUBCONTRACTORS, and corporations furnishing materials for or performing labor in the prosecution of the WORK provided for in such contract, and any authorized extension or modification thereof, including all amounts due for materials, lubricants, oil, gasoline, coal and coke, repairs on machinery, equipment and tools, consumed or used in connection with the construction of such WORK, and all insurance premiums on said WORK, and for all labor, performed in such WORK whether by SUBCONTRACTOR or otherwise, then this obligation shall be void; otherwise to remain in full force and effect.

CONTRACT DOCUMENTS FOR CONSTRUCTION OF
FEDERALLY ASSISTED WATER AND SEWER PROJECTS

Document No. 6
Payment Bond: Page 1 of 2

PROVIDED, FURTHER, that the said Surety for value received hereby stipulates and agrees that no change, extension of time, alteration or addition to the terms of the contract or to the WORK to be performed thereunder or the SPECIFICATIONS accompanying the same shall in any wise affect its obligation on this BOND, and it does hereby waive notice of any such change, extension of time, alteration or addition to the terms of the contract or to the WORK or to the SPECIFICATIONS.

PROVIDED, FURTHER, that no final settlement between the OWNER and the CONTRACTOR shall abridge the right of any beneficiary hereunder, whose claim may be unsatisfied.

IN WITNESS WHEREOF, this instrument is executed in _____(number) counterparts, each one of which shall be deemed an original, this the _____ day of _____ 19_____.

ATTEST:

_____
Principal

_____
(Principal) Secretary

(SEAL)   By _____(s)

_____
(Address)

_____

_____
Witness as to Principal

_____
(Address)

_____
Surety

ATTEST: By _____
Attorney-in-Fact

_____
(Address)

_____
Witness as to Surety

_____
(Address)

NOTE: Date of BOND must not be prior to date of Contract.
If CONTRACTOR is Partnership, all partners should execute BOND.

IMPORTANT: Surety companies executing BONDS must appear on the Treasury Department's most current list (Circular 570 as amended) and be authorized to transact business in the State where the PROJECT is located.

PERFORMANCE BOND

KNOW ALL MEN BY THESE PRESENTS: that

_____
(Name of Contractor)

_____
(Address of Contractor)

a _____, hereinafter called Principal, and
   (Corporation, Partnership, or Individual)

_____
(Name of Surety)

_____
(Address of Surety)

hereinafter called Surety, are held and firmly bound unto _____

_____
(Name of owner)

_____
(Address of Owner)

hereinafter called OWNER, in the penal sum of _____

_____Dollars, $(_____)

in lawful money of the United States, for the payment of which sum well and truly to be made, we bind ourselves, successors, and assigns, jointly and severally, firmly by these presents.

THE CONDITION OF THIS OBLIGATION is such that whereas, the Principal entered into a certain contract with the OWNER, dated the _____ day of_____, 19____, a copy of which is hereto attached and made a part hereof for the construction of:

_____

_____

_____

_____

NOW, THEREFORE, if the Principal shall well, truly and faithfully perform its duties, all the undertakings, covenants, terms, conditions, and agreements of said contract during the original term thereof, and any extensions thereof which may be granted by the OWNER, with or without notice to the Surety and during the one year guaranty period, and if he shall satisfy all claims and demands incurred under such contract, and shall fully indemnify and save harmless the OWNER from all costs and damages which it may suffer by reason of failure to do so, and shall reimburse and repay the OWNER all outlay and expense which the OWNER may incur in making good any default, then this obligation shall be void; otherwise to remain in full force and effect.

PROVIDED, FURTHER, that the said surety, for value received hereby stipulates and agrees that no change, extension of time, alteration or addition to the terms of the contract or to WORK to be performed thereunder or the SPECIFICATIONS accompanying the same shall in any wise affect its obligation on this BOND, and it does hereby waive notice of any such change, extension of time, alteration or addition to the terms of the contract or to the WORK or to the SPECIFICATIONS.

PROVIDED, FURTHER, that no final settlement between the OWNER and the CONTRACTOR shall abridge the right of any beneficiary hereunder, whose claim may be unsatisfied.

IN WITNESS WHEREOF, this instrument is executed in _____ counterparts, each
(Number)

one of which shall be deemed an original, this the _____ day of _____ 19_____.

ATTEST:

_____
(Principal) Secretary

(SEAL)

_____
(Witness as to Principal)

_____
(Address)

_____
Principal

By _____ (s)

_____
(Address)

ATTEST:

_____
(Surety) Secretary

(SEAL)

_____
Witness as to Surety

_____
(Address)

_____
Surety

By _____
Attorney-in-Fact

_____
(Address)

NOTE: Date of BOND must not be prior to date of Contract.
      If CONTRACTOR is Partnership, all partners should execute BOND.

IMPORTANT: Surety companies executing BONDS must appear on the Treasury Department's most current list (Circular 570 as amended) and be authorized to transact business in the state where the PROJECT is located.

Document No. 7
Performance Bond: Page 2 of 2

NOTICE OF AWARD

To: _____

_____

_____

_____

PROJECT Description: _____

_____

_____

    The OWNER has considered the BID submitted by you for the above described WORK in response to its Advertisement for Bids dated _____, 19 _____, and Information for Bidders.

    You are hereby notified that your BID has been accepted for items in the amount of $_____.

    You are required by the Information for Bidders to execute the Agreement and furnish the required CONTRACTOR'S Performance BOND, Payment BOND and certificates of insurance within ten (10) calendar days from the date of this Notice to you.

    If you fail to execute said Agreement and to furnish said BONDS within ten (10) days from the date of this Notice, said OWNER will be entitled to consider all your rights arising out of the OWNER'S acceptance of your BID as abandoned and as a forfeiture of your BID BOND. The OWNER will be entitled to such other rights as may be granted by law.

    You are required to return an acknowledged copy of this NOTICE OF AWARD to the OWNER.

Dated this _____ day of _____, 19_____.

_____
Owner

By _____

Title _____

ACCEPTANCE OF NOTICE

Receipt of the above NOTICE OF AWARD is hereby acknowledged

by _____,

this the _____ day of _____, 19_____

By _____

Title _____

CONTRACT DOCUMENTS FOR CONSTRUCTION OF
FEDERALLY ASSISTED WATER AND SEWER PROJECTS

Document No. 8
Notice of Award: Page 1 of 1

NOTICE TO PROCEED

To: _____   Date: _____

_____   Project: _____

_____   _____

_____   _____

_____

    You are hereby notified to commence WORK in accordance with the Agreement dated _____, 19_____, on or before _____, 19_____, and you are to complete the WORK within _____ consecutive calendar days thereafter. The date of completion of all WORK is therefore _____, 19_____.

_____
Owner

By _____

Title _____

ACCEPTANCE OF NOTICE

Receipt of the above NOTICE TO PROCEED is hereby acknowledged by _____

_____,

this the _____ day

of _____, 19_____

_____

By _____

Title _____

CONTRACT DOCUMENTS FOR CONSTRUCTION OF
FEDERALLY ASSISTED WATER AND SEWER PROJECTS

CHANGE ORDER

Order No. _____

Date: _____

Agreement Date: _____

NAME OF PROJECT: _____
_____

OWNER: _____

CONTRACTOR: _____

The following changes are hereby made to the CONTRACT DOCUMENTS:

Justification:

Change to CONTRACT PRICE:

Original CONTRACT PRICE $_____

Current CONTRACT PRICE adjusted by previous CHANGE ORDER $_____

The CONTRACT PRICE due to this CHANGE ORDER will be (increased) (decreased) by: $_____

The new CONTRACT PRICE including this CHANGE ORDER will be $_____

Change to CONTRACT TIME:

The CONTRACT TIME will be (increased) (decreased) by_____calendar days.

The date for completion of all work will be _____(Date).

Approvals Required:
To be effective this Order must be approved by the Federal agency if it changes the scope or objective of the PROJECT, or as may otherwise be required by the SUPPLEMENTAL GENERAL CONDITIONS.

Requested by: _____

Recommended by: _____

Ordered by: _____

Accepted by: _____

Federal Agency Approval (where applicable) _____

CONTRACT DOCUMENTS FOR CONSTRUCTION OF
FEDERALLY ASSISTED WATER AND SEWER PROJECTS

Document No. 10
Change Order: Page 1 of 1

GENERAL CONDITIONS

1. Definitions
2. Additional Instructions and Detail Drawings
3. Schedules, Reports and Records
4. Drawings and Specifications
5. Shop Drawings
6. Materials, Services and Facilities
7. Inspection and Testing
8. Substitutions
9. Patents
10. Surveys, Permits, Regulations
11. Protection of Work, Property, Persons
12. Supervision by Contractor
13. Changes in the Work
14. Changes in Contract Price
15. Time for Completion and Liquidated Damages
16. Correction of Work
17. Subsurface Conditions
18. Suspension of Work, Termination and Delay
19. Payments to Contractor
20. Acceptance of Final Payment as Release
21. Insurance
22. Contract Security
23. Assignments
24. Indemnification
25. Separate Contracts
26. Subcontracting
27. Engineer's Authority
28. Land and Rights-of-Way
29. Guaranty
30. Arbitration
31. Taxes

1. *DEFINITIONS*

1.1 Wherever used in the CONTRACT DOCUMENTS, the following terms shall have the meanings indicated which shall be applicable to both the singular and plural thereof:

1.2 ADDENDA—Written or graphic instruments issued prior to the execution of the Agreement which modify or interpret the CONTRACT DOCUMENTS, DRAWINGS and SPECIFICATIONS, by additions, deletions, clarifications or corrections.

1.3 BID—The offer or proposal of the BIDDER submitted on the prescribed form setting forth the prices for the WORK to be performed.

1.4 BIDDER—Any person, firm or corporation submitting a BID for the WORK.

1.5 BONDS—Bid, Performance, and Payment Bonds and other instruments of security, furnished by the CONTRACTOR and his surety in accordance with the CONTRACT DOCUMENTS.

1.6 CHANGE ORDER—A written order to the CONTRACTOR authorizing an addition, deletion or revision in the WORK within the general scope of the CONTRACT DOCUMENTS, or authorizing an adjustment in the CONTRACT PRICE or CONTRACT TIME.

1.7 CONTRACT DOCUMENTS—The contract, including Advertisement For Bids, Information For Bidders, BID, Bid Bond, Agreement, Payment Bond, Performance Bond, NOTICE OF AWARD, NOTICE TO PROCEED, CHANGE ORDER, DRAWINGS, SPECIFICATIONS, and ADDENDA.

1.8 CONTRACT PRICE—The total monies payable to the CONTRACTOR under the terms and conditions of the CONTRACT DOCUMENTS.

1.9 CONTRACT TIME—The number of calendar days stated in the CONTRACT DOCUMENTS for the completion of the WORK.

1.10 CONTRACTOR—The person, firm or corporation with whom the OWNER has executed the Agreement.

1.11 DRAWINGS—The part of the CONTRACT DOCUMENTS which show the characteristics and scope of the WORK to be performed and which have been prepared or approved by the ENGINEER.

1.12 ENGINEER—The person, firm or corporation named as such in the CONTRACT DOCUMENTS.

1.13 FIELD ORDER—A written order effecting a change in the WORK not involving an adjustment in the CONTRACT PRICE or an extension of the CONTRACT TIME, issued by the ENGINEER to the CONTRACTOR during construction.

1.14 NOTICE OF AWARD—The written notice of the acceptance of the BID from the OWNER to the successful BIDDER.

1.15 NOTICE TO PROCEED—Written communication issued by the OWNER to the CONTRACTOR authorizing him to proceed with the WORK and establishing the date of commencement of the WORK.

1.16 OWNER—A public or quasi-public body or authority, corporation, association, partnership, or individual for whom the WORK is to be performed.

1.17 PROJECT—The undertaking to be performed as provided in the CONTRACT DOCUMENTS.

1.18 RESIDENT PROJECT REPRESENTATIVE—The authorized representative of the OWNER who is assigned to the PROJECT site or any part thereof.

1.19 SHOP DRAWINGS—All drawings, diagrams, illustrations, brochures, schedules and other data which are prepared by the CONTRACTOR, a SUBCONTRACTOR, manufacturer, SUPPLIER or distributor, which illustrate how specific portions of the WORK shall be fabricated or installed.

1.20 SPECIFICATIONS—A part of the CONTRACT DOCUMENTS consisting of written descriptions of a technical nature of materials, equipment, construction systems, standards and workmanship.

1.21 SUBCONTRACTOR—An individual, firm or corporation having a direct contract with the CONTRACTOR or with any other SUBCONTRACTOR for the performance of a part of the WORK at the site.

1.22 SUBSTANTIAL COMPLETION—That date as certified by the ENGINEER when the construction of the PROJECT or a specified part thereof is sufficiently completed, in accordance with the CONTRACT DOCUMENTS, so that the PROJECT or specified part can be utilized for the purposes for which it is intended.

1.23 SUPPLEMENTAL GENERAL CONDITIONS—

Modifications to General Conditions required by a Federal agency for participation in the PROJECT and approved by the agency in writing prior to inclusion in the CONTRACT DOCUMENTS, or such requirements that may be imposed by applicable state laws.

1.24 SUPPLIER—Any person or organization who supplies materials or equipment for the WORK, including that fabricated to a special design, but who does not perform labor at the site.

1.25 WORK—All labor necessary to produce the construction required by the CONTRACT DOCUMENTS, and all materials and equipment incorporated or to be incorporated in the PROJECT.

1.26 WRITTEN NOTICE—Any notice to any party of the Agreement relative to any part of this Agreement in writing and considered delivered and the service thereof completed, when posted by certified or registered mail to the said party at his last given address, or delivered in person to said party or his authorized representative on the WORK.

2. ADDITIONAL INSTRUCTIONS AND DETAIL DRAWINGS

2.1 The CONTRACTOR may be furnished additional instructions and detail drawings, by the ENGINEER, as necessary to carry out the WORK required by the CONTRACT DOCUMENTS.

2.2 The additional drawings and instruction thus supplied will become a part of the CONTRACT DOCUMENTS. The CONTRACTOR shall carry out the WORK in accordance with the additional detail drawings and instructions.

3. SCHEDULES, REPORTS AND RECORDS

3.1 The CONTRACTOR shall submit to the OWNER such schedule of quantities and costs, progress schedules, payrolls, reports, estimates, records and other data where applicable as are required by the CONTRACT DOCUMENTS for the WORK to be performed.

3.2 Prior to the first partial payment estimate the CONTRACTOR shall submit construction progress schedules showing the order in which he proposes to carry on the WORK, including dates at which he will start the various parts of the WORK, estimated date of completion of each part and, as applicable:

3.2.1 The dates at which special detail drawings will be required; and

3.2.2 Respective dates for submission of SHOP DRAWINGS, the beginning of manufacture, the testing and the installation of materials, supplies and equipment.

3.3 The CONTRACTOR shall also submit a schedule of payments that he anticipates he will earn during the course of the WORK.

4. DRAWINGS AND SPECIFICATIONS

4.1 The intent of the DRAWINGS and SPECIFICATIONS is that the CONTRACTOR shall furnish all labor, materials, tools, equipment, and transportation necessary for the proper execution of the WORK in accordance with the CONTRACT DOCUMENTS and all incidental work necessary to complete the PROJECT in an acceptable manner, ready for use, occupancy or operation by the OWNER.

4.2 In case of conflict between the DRAWINGS and SPECIFICATIONS, the SPECIFICATIONS shall govern. Figure dimensions on DRAWINGS shall govern over scale dimensions, and detailed DRAWINGS shall govern over general DRAWINGS.

4.3 Any discrepancies found between the DRAWINGS and SPECIFICATIONS and site conditions or any inconsistencies or ambiguities in the DRAWINGS or SPECIFICATIONS shall be immediately reported to the ENGINEER, in writing, who shall promptly correct such inconsistencies or ambiguities in writing. WORK done by the CONTRACTOR after his discovery of such discrepancies, inconsistencies or ambiguities shall be done at the CONTRACTOR'S risk.

5. SHOP DRAWINGS

5.1 The CONTRACTOR shall provide SHOP DRAWINGS as may be necessary for the prosecution of the WORK as required by the CONTRACT DOCUMENTS. The ENGINEER shall promptly review all SHOP DRAWINGS. The ENGINEER'S approval of any SHOP DRAWING shall not release the CONTRACTOR from responsibility for deviations from the CONTRACT DOCUMENTS. The approval of any SHOP DRAWING which substantially deviates from the requirement of the CONTRACT DOCUMENTS shall be evidenced by a CHANGE ORDER.

5.2 When submitted for the ENGINEER'S review, SHOP DRAWINGS shall bear the CONTRACTOR'S certification that he has reviewed, checked and approved the SHOP DRAWINGS and that they are in conformance with the requirements of the CONTRACT DOCUMENTS.

5.3 Portions of the WORK requiring a SHOP DRAWING or sample submission shall not begin until the SHOP DRAWING or submission has been approved by the ENGINEER. A copy of each approved SHOP DRAWING and each approved sample shall be kept in good order by the CONTRACTOR at the site and shall be available to the ENGINEER.

6. MATERIALS, SERVICES AND FACILITIES

6.1 It is understood that, except as otherwise specifically stated in the CONTRACT DOCUMENTS, the CONTRACTOR shall provide and pay for all materials, labor, tools, equipment, water, light, power, transportation, supervision, temporary construction of any nature, and all other services and facilities of any nature whatsoever necessary to execute, complete, and deliver the WORK within the specified time.

6.2 Materials and equipment shall be so stored as to insure the preservation of their quality and fitness for the WORK. Stored materials and equipment to be incorporated in the WORK shall be located so as to facilitate prompt inspection.

6.3 Manufactured articles, materials and equipment shall be applied, installed, connected, erected, used, cleaned and conditioned as directed by the manufacturer.

6.4 Materials, supplies and equipment shall be in accordance with samples submitted by the CONTRACTOR and approved by the ENGINEER.

6.5 Materials, supplies or equipment to be incorporated into the WORK shall not be purchased by the

CONTRACTOR or the SUBCONTRACTOR subject to a chattel mortgage or under a conditional sale contract or other agreement by which an interest is retained by the seller.

7. INSPECTION AND TESTING

7.1 All materials and equipment used in the construction of the PROJECT shall be subject to adequate inspection and testing in accordance with generally accepted standards, as required and defined in the CONTRACT DOCUMENTS.

7.2 The OWNER shall provide all inspection and testing services not required by the CONTRACT DOCUMENTS.

7.3 The CONTRACTOR shall provide at his expense the testing and inspection services required by the CONTRACT DOCUMENTS.

7.4 If the CONTRACT DOCUMENTS, laws, ordinances, rules, regulations or orders of any public authority having jurisdiction require any WORK to specifically be inspected, tested, or approved by someone other than the CONTRACTOR, the CONTRACTOR will give the ENGINEER timely notice of readiness. The CONTRACTOR will then furnish the ENGINEER the required certificates of inspection, testing or approval.

7.5 Inspections, tests or approvals by the engineer or others shall not relieve the CONTRACTOR from his obligations to perform the WORK in accordance with the requirements of the CONTRACT DOCUMENTS.

7.6 The ENGINEER and his representatives will at all times have access to the WORK. In addition, authorized representatives and agents of any participating Federal or state agency shall be permitted to inspect all work, materials, payrolls, records of personnel, invoices of materials, and other relevant data and records. The CONTRACTOR will provide proper facilities for such access and observation of the WORK and also for any inspection, or testing thereof.

7.7 If any WORK is covered contrary to the written instructions of the ENGINEER it must, if requested by the ENGINEER, be uncovered for his observation and replaced at the CONTRACTOR'S expense.

7.8 If the ENGINEER considers it necessary or advisable that covered WORK be inspected or tested by others, the CONTRACTOR, at the ENGINEER'S request, will uncover, expose or otherwise make available for observation, inspection or testing as the ENGINEER may require, that portion of the WORK in question, furnishing all necessary labor, materials, tools, and equipment. If it is found that such WORK is defective, the CONTRACTOR will bear all the expenses of such uncovering, exposure, observation, inspection and testing and of satisfactory reconstruction. If, however, such WORK is not found to be defective, the CONTRACTOR will be allowed an increase in the CONTRACT PRICE or an extension of the CONTRACT TIME, or both, directly attributable to such uncovering, exposure, observation, inspection, testing and reconstruction and an appropriate CHANGE ORDER shall be issued.

8. SUBSTITUTIONS

8.1 Whenever a material, article or piece of equipment is identified on the DRAWINGS or SPECIFICATIONS by reference to brand name or catalogue number, it shall be understood that this is referenced for the purpose of defining the performance or other salient requirements and that other products of equal capacities, quality and function shall be considered. The CONTRACTOR may recommend the substitution of a material, article, or piece of equipment of equal substance and function for those referred to in the CONTRACT DOCUMENTS by reference to brand name or catalogue number, and if, in the opinion of the ENGINEER, such material, article, or piece of equipment is of equal substance and function to that specified, the ENGINEER may approve its substitution and use by the CONTRACTOR. Any cost differential shall be deductible from the CONTRACT PRICE and the CONTRACT DOCUMENTS shall be appropriately modified by CHANGE ORDER. The CONTRACTOR warrants that if substitutes are approved, no major changes in the function or general design of the PROJECT will result. Incidental changes or extra component parts required to accommodate the substitute will be made by the CONTRACTOR without a change in the CONTRACT PRICE or CONTRACT TIME.

9. PATENTS

9.1 The CONTRACTOR shall pay all applicable royalties and license fees. He shall defend all suits or claims for infringement of any patent rights and save the OWNER harmless from loss on account thereof, except that the OWNER shall be responsible for any such loss when a particular process, design, or the product of a particular manufacturer or manufacturers is specified, however if the CONTRACTOR has reason to believe that the design, process or product specified is an infringement of a patent, he shall be responsible for such loss unless he promptly gives such information to the ENGINEER.

10. SURVEYS, PERMITS, REGULATIONS

10.1 The OWNER shall furnish all boundary surveys and establish all base lines for locating the principal component parts of the WORK together with a suitable number of bench marks adjacent to the WORK as shown in the CONTRACT DOCUMENTS. From the information provided by the OWNER, unless otherwise specified in the CONTRACT DOCUMENTS, the CONTRACTOR shall develop and make all detail surveys needed for construction such as slope stakes, batter boards, stakes for pile locations and other working points, lines, elevations and cut sheets.

10.2 The CONTRACTOR shall carefully preserve bench marks, reference points and stakes and, in case of willful or careless destruction, he shall be charged with the resulting expense and shall be responsible for any mistakes that may be caused by their unnecessary loss or disturbance.

10.3 Permits and licenses of a temporary nature necessary for the prosecution of the WORK shall be secured and paid for by the CONTRACTOR unless otherwise stated in the SUPPLEMENTAL GENERAL CONDITIONS. Permits, licenses and easements for permanent structures or permanent changes in existing facilities shall be secured and paid for by the OWNER, unless otherwise specified. The CONTRACTOR shall give all notices and comply with all laws, ordinances, rules and regulations bearing on the conduct of the WORK as drawn and specified. If the CONTRACTOR observes that the CONTRACT DOCUMENTS are at variance therewith, he shall promptly notify the ENGINEER in writing, and any necessary changes shall be adjusted as provided in Section 13, CHANGES IN THE WORK.

11. PROTECTION OF WORK, PROPERTY AND PERSONS

11.1 The CONTRACTOR will be responsible for initiating, maintaining and supervising all safety precautions and programs in connection with the WORK. He will take all necessary precautions for the safety of, and will provide the necessary protection to prevent damage, injury or loss to all employees on the WORK and other persons who may be affected thereby, all the WORK and all materials or equipment to be incorporated therein, whether in storage on or off the site, and other property at the site or adjacent thereto, including trees, shrubs, lawns, walks, pavements, roadways, structures and utilities not designated for removal, relocation or replacement in the course of construction.

11.2 The CONTRACTOR will comply with all applicable laws, ordinances, rules, regulations and orders of any public body having jurisdiction. He will erect and maintain, as required by the conditions and progress of the WORK, all necessary safeguards for safety and protection. He will notify owners of adjacent utilities when prosecution of the WORK may affect them. The CONTRACTOR will remedy all damage, injury or loss to any property caused, directly or indirectly, in whole or in part, by the CONTRACTOR, any SUBCONTRACTOR or anyone directly or indirectly employed by any of them or anyone for whose acts any of them be liable, except damage or loss attributable to the fault of the CONTRACT DOCUMENTS or to the acts or omissions of the OWNER or the ENGINEER or anyone employed by either of them or anyone for whose acts either of them may be liable, and not attributable, directly or indirectly, in whole or in part, to the fault or negligence of the CONTRACTOR.

11.3 In emergencies affecting the safety of persons or the WORK or property at the site or adjacent thereto, the CONTRACTOR, without special instruction or authorization from the ENGINEER or OWNER, shall act to prevent threatened damage, injury or loss. He will give the ENGINEER prompt WRITTEN NOTICE of any significant changes in the WORK or deviations from the CONTRACT DOCUMENTS caused thereby, and a CHANGE ORDER shall thereupon be issued covering the changes and deviations involved.

12. SUPERVISION BY CONTRACTOR

12.1 The CONTRACTOR will supervise and direct the WORK. He will be solely responsible for the means, methods, techniques, sequences and procedures of construction. The CONTRACTOR will employ and maintain on the WORK a qualified supervisor or superintendent who shall have been designated in writing by the CONTRACTOR as the CONTRACTOR'S representative at the site. The supervisor shall have full authority to act on behalf of the CONTRACTOR and all communications given to the supervisor shall be as binding as if given to the CONTRACTOR. The supervisor shall be present on the site at all times as required to perform adequate supervision and coordination of the WORK.

13. CHANGES IN THE WORK

13.1 The OWNER may at any time, as the need arises, order changes within the scope of the WORK without invalidating the Agreement. If such changes increase or decrease the amount due under the CONTRACT DOCUMENTS, or in the time required for performance of the WORK, an equitable adjustment shall be authorized by CHANGE ORDER.

13.2 The ENGINEER, also, may at any time, by issuing a FIELD ORDER, make changes in the details of the WORK. The CONTRACTOR shall proceed with the performance of any changes in the WORK so ordered by the ENGINEER unless the CONTRACTOR believes that such FIELD ORDER entitles him to a change in CONTRACT PRICE or TIME, or both, in which event he shall give the ENGINEER WRITTEN NOTICE thereof within seven (7) days after the receipt of the ordered change. Thereafter the CONTRACTOR shall document the basis for the change in CONTRACT PRICE or TIME within thirty (30) days. The CONTRACTOR shall not execute such changes pending the receipt of an executed CHANGE ORDER or further instruction from the OWNER.

14. CHANGES IN CONTRACT PRICE

14.1 The CONTRACT PRICE may be changed only by a CHANGE ORDER. The value of any WORK covered by a CHANGE ORDER or of any claim for increase or decrease in the CONTRACT PRICE shall be determined by one or more of the following methods in the order of precedence listed below:
(a) Unit prices previously approved.
(b) An agreed lump sum.
(c) The actual cost for labor, direct overhead, materials, supplies, equipment, and other services necessary to complete the work. In addition there shall be added an amount to be agreed upon but not to exceed fifteen (15) percent of the actual cost of the WORK to cover the cost of general overhead and profit.

15. TIME FOR COMPLETION AND LIQUIDATED DAMAGES

15.1 The date of beginning and the time for completion of the WORK are essential conditions of the CONTRACT DOCUMENTS and the WORK embraced shall be commenced on a date specified in the NOTICE TO PROCEED.

15.2 The CONTRACTOR will proceed with the WORK at such rate of progress to insure full completion within the CONTRACT TIME. It is expressly understood and agreed, by and between the CONTRACTOR and the OWNER, that the CONTRACT TIME for the completion of the WORK described herein is a reasonable time, taking into consideration the average climatic and economic conditions and other factors prevailing in the locality of the WORK.

15.3 If the CONTRACTOR shall fail to complete the WORK within the CONTRACT TIME, or extension of time granted by the OWNER, then the CONTRACTOR will pay to the OWNER the amount for liquidated damages as specified in the BID for each calendar day that the CONTRACTOR shall be in default after the time stipulated in the CONTRACT DOCUMENTS.

15.4 The CONTRACTOR shall not be charged with liquidated damages or any excess cost when the delay in completion of the WORK is due to the following, and the CONTRACTOR has promptly given WRITTEN NOTICE of such delay to the OWNER or ENGINEER.

15.4.1 To any preference, priority or allocation

order duly issued by the OWNER.

15.4.2 To unforeseeable causes beyond the control and without the fault or negligence of the CONTRACTOR, including but not restricted to, acts of God, or of the public enemy, acts of the OWNER, acts of another CONTRACTOR in the performance of a contract with the OWNER, fires, floods, epidemics, quarantine restrictions, strikes, freight embargoes, and abnormal and unforeseeable weather; and

15.4.3 To any delays of SUBCONTRACTORS occasioned by any of the causes specified in paragraphs 15.4.1 and 15.4.2 of this article.

16. CORRECTION OF WORK

16.1 The CONTRACTOR shall promptly remove from the premises all WORK rejected by the ENGINEER for failure to comply with the CONTRACT DOCUMENTS, whether incorporated in the construction or not, and the CONTRACTOR shall promptly replace and re-execute the WORK in accordance with the CONTRACT DOCUMENTS and without expense to the OWNER and shall bear the expense of making good all WORK of other CONTRACTORS destroyed or damaged by such removal or replacement.

16.2 All removal and replacement WORK shall be done at the CONTRACTOR'S expense. If the CONTRACTOR does not take action to remove such rejected WORK within ten (10) days after receipt of WRITTEN NOTICE, the OWNER may remove such WORK and store the materials at the expense of the CONTRACTOR.

17. SUBSURFACE CONDITIONS

17.1 The CONTRACTOR shall promptly, and before such conditions are disturbed, except in the event of an emergency, notify the OWNER by WRITTEN NOTICE of:

17.1.1 Subsurface or latent physical conditions at the site differing materially from those indicated in the CONTRACT DOCUMENTS; or

17.1.2 Unknown physical conditions at the site, of an unusual nature, differing materially from those ordinarily encountered and generally recognized as inherent in WORK of the character provided for in the CONTRACT DOCUMENTS.

17.2 The OWNER shall promptly investigate the conditions, and if he finds that such conditions do so materially differ and cause an increase or decrease in the cost of, or in the time required for, performance of the WORK, an equitable adjustment shall be made and the CONTRACT DOCUMENTS shall be modified by a CHANGE ORDER. Any claim of the CONTRACTOR for adjustment hereunder shall not be allowed unless he has given the required WRITTEN NOTICE; provided that the OWNER may, if he determines the facts so justify, consider and adjust any such claims asserted before the date of final payment.

18. SUSPENSION OF WORK, TERMINATION AND DELAY

18.1 The OWNER may suspend the WORK or any portion thereof for a period of not more than ninety days or such further time as agreed upon by the CONTRACTOR, by WRITTEN NOTICE to the CONTRACTOR and the ENGINEER which notice shall fix the date on which WORK shall be resumed. The CONTRACTOR will resume that WORK on the date so fixed. The CONTRACTOR will be allowed an increase in the CONTRACT PRICE or an extension of the CONTRACT TIME, or both, directly attributable to any suspension.

18.2 If the CONTRACTOR is adjudged a bankrupt or insolvent, or if he makes a general assignment for the benefit of his creditors, or if a trustee or receiver is appointed for the CONTRACTOR or for any of his property, or if he files a petition to take advantage of any debtor's act, or to reorganize under the bankruptcy or applicable laws, or if he repeatedly fails to supply sufficient skilled workmen or suitable materials or equipment, or if he repeatedly fails to make prompt payments to SUBCONTRACTORS or for labor, materials or equipment or if he disregards laws, ordinances, rules, regulations or orders of any public body having jurisdiction of the WORK or if he disregards the authority of the ENGINEER, or if he otherwise violates any provision of the CONTRACT DOCUMENTS, then the OWNER may, without prejudice to any other right or remedy and after giving the CONTRACTOR and his surety a minimum of ten (10) days from delivery of a WRITTEN NOTICE, terminate the services of the CONTRACTOR and take possession of the PROJECT and of all materials, equipment, tools, construction equipment and machinery thereon owned by the CONTRACTOR, and finish the WORK by whatever method he may deem expedient. In such case the CONTRACTOR shall not be entitled to receive any further payment until the WORK is finished. If the unpaid balance of the CONTRACT PRICE exceeds the direct and indirect costs of completing the PROJECT, including compensation for additional professional services, such excess SHALL BE PAID TO THE CONTRACTOR. If such costs exceed such unpaid balance, the CONTRACTOR will pay the difference to the OWNER. Such costs incurred by the OWNER will be determined by the ENGINEER and incorporated in a CHANGE ORDER.

18.3 Where the CONTRACTOR'S services have been so terminated by the OWNER, said termination shall not affect any right of the OWNER against the CONTRACTOR then existing or which may thereafter accrue. Any retention or payment of monies by the OWNER due the CONTRACTOR will not release the CONTRACTOR from compliance with the CONTRACT DOCUMENTS.

18.4 After ten (10) days from delivery of a WRITTEN NOTICE to the CONTRACTOR and the ENGINEER, the OWNER may, without cause and without prejudice to any other right or remedy, elect to abandon the PROJECT and terminate the Contract. In such case, the CONTRACTOR shall be paid for all WORK executed and any expense sustained plus reasonable profit.

18.5 If, through no act or fault of the CONTRACTOR, the WORK is suspended for a period of more than ninety (90) days by the OWNER or under an order of court or other public authority, or the ENGINEER fails to act on any request for payment within thirty (30) days after it is submitted, or the OWNER fails to pay the CONTRACTOR substantially the sum approved by the ENGINEER or awarded by arbitrators within thirty (30) days of its approval and presentation, then the CONTRACTOR may, after ten (10) days from delivery of a WRITTEN NOTICE to the OWNER and the ENGINEER, terminate the CONTRACT and recover from the OWNER payment for all WORK executed and all expenses sustained. In addition and in lieu of terminating the CONTRACT, if the ENGINEER has failed to act on a request for payment or if the OWNER has failed to make any payment as aforesaid, the CONTRACTOR may upon ten (10) days written notice to the OWNER and the ENGINEER stop the WORK until he has been paid all amounts then due, in which event and upon resumption of the WORK, CHANGE ORDERS shall be issued for adjusting the CONTRACT PRICE or extending the CONTRACT TIME or both to compensate for the costs and delays attributable to the stoppage of the WORK.

18.6 If the performance of all or any portion of the WORK is suspended, delayed, or interrupted as a result of a failure of the OWNER or ENGINEER to act within the time specified in the CONTRACT DOCUMENTS, or if no time is specified, within a reasonable time, an adjustment in the CONTRACT PRICE or an extension of the CONTRACT TIME, or both, shall be made by CHANGE ORDER to compensate the CONTRACTOR for the costs and delays necessarily caused by the failure of the OWNER or ENGINEER.

19. PAYMENTS TO CONTRACTOR

19.1 At least ten (10) days before each progress payment falls due (but not more often than once a month), the CONTRACTOR will submit to the ENGINEER a partial payment estimate filled out and signed by the CONTRACTOR covering the WORK performed during the period covered by the partial payment estimate and supported by such data as the ENGINEER may reasonably require. If payment is requested on the basis of materials and equipment not incorporated in the WORK but delivered and suitably stored at or near the site, the partial payment estimate shall also be accompanied by such supporting data, satisfactory to the OWNER, as will establish the OWNER's title to the material and equipment and protect his interest therein, including applicable insurance. The ENGINEER will, within ten (10) days after receipt of each partial payment estimate, either indicate in writing his approval of payment and present the partial payment estimate to the OWNER, or return the partial payment estimate to the CONTRACTOR indicating in writing his reasons for refusing to approve payment. In the latter case, the CONTRACTOR may make the necessary corrections and resubmit the partial payment estimate. The OWNER will, within ten (10) days of presentation to him of an approved partial payment estimate, pay the CONTRACTOR a progress payment on the basis of the approved partial payment estimate. The OWNER shall retain ten (10) percent of the amount of each payment until final completion and acceptance of all work covered by the CONTRACT DOCUMENTS. The OWNER at any time, however, after fifty (50) percent of the WORK has been completed, if he finds that satisfactory progress is being made, shall reduce retainage to five (5%) percent on the current and remaining estimates. When the WORK is substantially complete (operational or beneficial occupancy), the retained amount may be further reduced below five (5) percent to only that amount necessary to assure completion. On completion and acceptance of a part of the WORK on which the price is stated separately in the CONTRACT DOCUMENTS, payment may be made in full, including retained percentages, less authorized deductions.

19.2 The request for payment may also include an allowance for the cost of such major materials and equipment which are suitably stored either at or near the site.

19.3 Prior to SUBSTANTIAL COMPLETION, the OWNER, with the approval of the ENGINEER and with the concurrence of the CONTRACTOR, may use any completed or substantially completed portions of the WORK. Such use shall not constitute an acceptance of such portions of the WORK.

19.4 The OWNER shall have the right to enter the premises for the purpose of doing work not covered by the CONTRACT DOCUMENTS. This provision shall not be construed as relieving the CONTRACTOR of the sole responsibility for the care and protection of the WORK, or the restoration of any damaged WORK except such as may be caused by agents or employees of the OWNER.

19.5 Upon completion and acceptance of the WORK, the ENGINEER shall issue a certificate attached to the final payment request that the WORK has been accepted by him under the conditions of the CONTRACT DOCUMENTS. The entire balance found to be due the CONTRACTOR, including the retained percentages, but except such sums as may be lawfully retained by the OWNER, shall be paid to the CONTRACTOR within thirty (30) days of completion and acceptance of the WORK.

19.6 The CONTRACTOR will indemnify and save the OWNER or the OWNER'S agents harmless from all claims growing out of the lawful demands of SUBCONTRACTORS, laborers, workmen, mechanics, materialmen, and furnishers of machinery and parts thereof, equipment, tools, and all supplies, incurred in the furtherance of the performance of the WORK. The CONTRACTOR shall, at the OWNER'S request, furnish satisfactory evidence that all obligations of the nature designated above have been paid, discharged, or waived. If the CONTRACTOR fails to do so the OWNER may, after having notified the CONTRACTOR, either pay unpaid bills or withhold from the CONTRACTOR'S unpaid compensation a sum of money deemed reasonably sufficient to pay any and all such lawful claims until satisfactory evidence is furnished that all liabilities have been fully discharged whereupon payment to the CONTRACTOR shall be resumed, in accordance with the terms of the CONTRACT DOCUMENTS, but in no event shall the provisions of this sentence be construed to impose any obligations upon the OWNER to either the CONTRACTOR, his Surety, or any third party. In paying any unpaid bills of the CONTRACTOR, any payment so made by the OWNER shall be considered as a payment made under the CONTRACT DOCUMENTS by the OWNER to the CONTRACTOR and the OWNER shall not be liable to the CONTRACTOR for any such payments made in good faith.

19.7 If the OWNER fails to make payment thirty (30) days after approval by the ENGINEER, in addition to other remedies available to the CONTRACTOR, there shall be added to each such payment interest at the maximum legal rate commencing on the first day after said payment is due and continuing until the payment is received by the CONTRACTOR.

20. *ACCEPTANCE OF FINAL PAYMENT AS RELEASE*

20.1 The acceptance by the CONTRACTOR of final payment shall be and shall operate as a release to the OWNER of all claims and all liability to the CONTRACTOR other than claims in stated amounts as may be specifically excepted by the CONTRACTOR for all things done or furnished in connection with this WORK and for every act and neglect of the OWNER and others relating to or arising out of this WORK. Any payment, however, final or otherwise, shall not release the CONTRACTOR or his sureties from any obligations under the CONTRACT DOCUMENTS or the Performance BOND and Payment BONDS.

21. *INSURANCE*

21.1 The CONTRACTOR shall purchase and maintain such insurance as will protect him from claims set forth below which may arise out of or result from the CONTRACTOR'S execution of the WORK, whether such execution be by himself or by any SUBCONTRACTOR or by anyone directly or indirectly employed by any of them, or by anyone for whose acts any of them may be liable:

21.1.1 Claims under workmen's compensation, disability benefit and other similar employee benefit acts;

21.1.2 Claims for damages because of bodily injury, occupational sickness or disease, or death of his employees;

21.1.3 Claims for damages because of bodily injury, sickness or disease, or death of any person other than his employees;

21.1.4 Claims for damages insured by usual personal injury liability coverage which are sustained (1) by any person as a result of an offense directly or indirectly related to the employment of such person by the CONTRACTOR, or (2) by any other person; and

21.1.5 Claims for damages because of injury to or destruction of tangible property, including loss of use resulting therefrom.

21.2 Certificates of Insurance acceptable to the OWNER shall be filed with the OWNER prior to commencement of the WORK. These Certificates shall contain a provision that coverages afforded under the policies will not be cancelled unless at least fifteen (15) days prior WRITTEN NOTICE has been given to the OWNER.

21.3 The CONTRACTOR shall procure and maintain, at his own expense, during the CONTRACT TIME, liability insurance as hereinafter specified;

21.3.1 CONTRACTOR'S General Public Liability and Property Damage Insurance including vehicle coverage issued to the CONTRACTOR and protecting him from all claims for personal injury, including death, and all claims for destruction of or damage to property, arising out of or in connection with any operations under the CONTRACT DOCUMENTS, whether such operations be by himself or by any SUBCONTRACTOR under him, or anyone directly or indirectly employed by the CONTRACTOR or by a SUBCONTRACTOR under him. Insurance shall be written with a limit of liability of not less than $500,000 for all damages arising out of bodily injury, including death, at any time resulting therefrom, sustained by any one person in any one accident; and a limit of liability of not less than $500,000 aggregate for any such damages sustained by two or more persons in any one accident. Insurance shall be written with a limit of liability of not less than $200,000 for all property damage sustained by any one person in any one accident; and a limit of liability of not less than $200,000 aggregate for any such damage sustained by two or more persons in any one accident.

21.3.2 The CONTRACTOR shall acquire and maintain, if applicable, Fire and Extended Coverage insurance upon the PROJECT to the full insurable value thereof for the benefit of the OWNER, the CONTRACTOR, and SUBCONTRACTORS as their interest may appear. This provision shall in no way release the CONTRACTOR or CONTRACTOR'S surety from obligations under the CONTRACT DOCUMENTS to fully complete the PROJECT.

21.4 The CONTRACTOR shall procure and maintain, at his own expense, during the CONTRACT TIME, in accordance with the provisions of the laws of the state in which the work is performed, Workmen's Compensation Insurance, including occupational disease provisions, for all of his employees at the site of the PROJECT and in case any work is sublet, the CONTRACTOR shall require such SUBCONTRACTOR similarly to provide Workmen's Compensation Insurance, including occupational disease provisions for all of the latter's employees unless such employees are covered by the protection afforded by the CONTRACTOR. In case any class of employees engaged in hazardous work under this contract at the site of the PROJECT is not protected under Workmen's Compensation statute, the CONTRACTOR shall provide, and shall cause each SUBCONTRACTOR to provide, adequate and suitable insurance for the protection of his employees not otherwise protected.

21.5 The CONTRACTOR shall secure, if applicable, "All Risk" type Builder's Risk Insurance for WORK to be performed. Unless specifically authorized by the OWNER, the amount of such insurance shall not be less than the CONTRACT PRICE totaled in the BID. The policy shall cover not less than the losses due to fire, explosion, hail, lightning, vandalism, malicious mischief, wind, collapse, riot, aircraft, and smoke during the CONTRACT TIME, and until the WORK is accepted by the OWNER. The policy shall name as the insured the CONTRACTOR, the ENGINEER, and the OWNER.

22. *CONTRACT SECURITY*

22.1 The CONTRACTOR shall within ten (10) days after the receipt of the NOTICE OF AWARD furnish the OWNER with a Performance Bond and a Payment Bond in penal sums equal to the amount of the CONTRACT PRICE, conditioned upon the performance by the CONTRACTOR of all undertakings, covenants, terms, conditions and agreements of the CONTRACT DOCUMENTS, and upon the prompt payment by the CONTRACTOR to all persons supplying labor and materials in the prosecution of the WORK provided by the CONTRACT DOCUMENTS. Such BONDS shall be executed by the CONTRACTOR and a corporate bonding company licensed to transact such business in the state in which the WORK is to be performed and named on the current list of "Surety Companies Acceptable on Federal Bonds" as published in the Treasury Department Circular Number 570. The expense of these BONDS shall be borne by the CONTRACTOR. If at any time a surety on any such BOND is declared a bankrupt or loses its right to do business in the state in which the WORK is to be performed or is removed from the list of Surety Companies accepted on Federal BONDS, CONTRACTOR shall within ten (10) days after notice from the OWNER to do so, substitute an acceptable BOND (or BONDS) in such form and sum and signed by such other surety or sureties as may be satisfactory to the OWNER. The premiums on such BOND shall be paid by the CONTRACTOR. No further payments shall be deemed due nor shall be made until the new surety or sureties shall have furnished an acceptable BOND to the OWNER.

23. ASSIGNMENTS

23.1 Neither the CONTRACTOR nor the OWNER shall sell, transfer, assign or otherwise dispose of the Contract or any portion thereof, or of his right, title or interest therein, or his obligations thereunder, without written consent of the other party.

24. INDEMNIFICATION

24.1 The CONTRACTOR will indemnify and hold harmless the OWNER and the ENGINEER and their agents and employees from and against all claims, damages, losses and expenses including attorney's fees arising out of or resulting from the performance of the WORK, provided that any such claims, damage, loss or expense is attributable to bodily injury, sickness, disease or death, or to injury to or destruction of tangible property including the loss of use resulting therefrom; and is caused in whole or in part by any negligent or willful act or omission of the CONTRACTOR, and SUBCONTRACTOR, anyone directly or indirectly employed by any of them or anyone for whose acts any of them may be liable.

24.2 In any and all claims against the OWNER or the ENGINEER, or any of their agents or employees, by any employee of the CONTRACTOR, any SUBCONTRACTOR, anyone directly or indirectly employed by any of them, or anyone for whose acts any of them may be liable, the indemnification obligation shall not be limited in any way by any limitation on the amount or type of damages, compensation or benefits payable by or for the CONTRACTOR or any SUBCONTRACTOR under workmen's compensation acts, disability benefit acts or other employee benefits acts.

24.3 The obligation of the CONTRACTOR under this paragraph shall not extend to the liability of the ENGINEER, his agents or employees arising out of the preparation or approval of maps, DRAWINGS, opinions, reports, surveys, CHANGE ORDERS, designs or SPECIFICATIONS.

25. SEPARATE CONTRACTS

25.1 The OWNER reserves the right to let other contracts in connection with this PROJECT. The CONTRACTOR shall afford other CONTRACTORS reasonable opportunity for the introduction and storage of their materials and the execution of their WORK, and shall properly connect and coordinate his WORK with theirs. If the proper execution or results of any part of the CONTRACTOR'S WORK depends upon the WORK of any other CONTRACTOR, the CONTRACTOR shall inspect and promptly report to the ENGINEER any defects in such WORK that render it unsuitable for such proper execution and results.

25.2 The OWNER may perform additional WORK related to the PROJECT by himself, or he may let other contracts containing provisions similar to these. The CONTRACTOR will afford the other CONTRACTORS who are parties to such Contracts (or the OWNER, if he is performing the additional WORK himself), reasonable opportunity for the introduction and storage of materials and equipment and the execution of WORK, and shall properly connect and coordinate his WORK with theirs.

25.3 If the performance of additional WORK by other CONTRACTORS or the OWNER is not noted in the CONTRACT DOCUMENTS prior to the execution of the CONTRACT, written notice thereof shall be given to the CONTRACTOR prior to starting any such additional WORK. If the CONTRACTOR believes that the performance of such additional WORK by the OWNER or others involves him in additional expense or entities him to an extension of the CONTRACT TIME, he may make a claim therefor as provided in Sections 14 and 15.

26. SUBCONTRACTING

26.1 The CONTRACTOR may utilize the services of specialty SUBCONTRACTORS on those parts of the WORK which, under normal contracting practices, are performed by specialty SUBCONTRACTORS.

26.2 The CONTRACTOR shall not award WORK to SUBCONTRACTOR(s), in excess of fifty (50%) percent of the CONTRACT PRICE, without prior written approval of the OWNER.

26.3 The CONTRACTOR shall be fully responsible to the OWNER for the acts and omissions of his SUBCONTRACTORS, and of persons either directly or indirectly employed by them, as he is for the acts and omissions of persons directly employed by him.

26.4 The CONTRACTOR shall cause appropriate provisions to be inserted in all subcontracts relative to the WORK to bind SUBCONTRACTORS to the CONTRACTOR by the terms of the CONTRACT DOCUMENTS insofar as applicable to the WORK of SUBCONTRACTORS and to give the CONTRACTOR the same power as regards terminating any subcontract that the OWNER may exercise over the CONTRACTOR under any provision of the CONTRACT DOCUMENTS.

26.5 Nothing contained in this CONTRACT shall create any contractual relation between any SUBCONTRACTOR and the OWNER.

27. ENGINEER'S AUTHORITY

27.1 The ENGINEER shall act as the OWNER'S representative during the construction period. He shall decide questions which may arise as to quality and acceptability of materials furnished and WORK performed. He shall interpret the intent of the CONTRACT DOCUMENTS in a fair and unbiased manner. The

ENGINEER will make visits to the site and determine if the WORK is proceeding in accordance with the CONTRACT DOCUMENTS.

27.2 The CONTRACTOR will be held strictly to the intent of the CONTRACT DOCUMENTS in regard to the quality of materials, workmanship and execution of the WORK. Inspections may be made at the factory or fabrication plant of the source of material supply.

27.3 The ENGINEER will not be responsible for the construction means, controls, techniques, sequences, procedures, or construction safety.

27.4 The ENGINEER shall promptly make decisions relative to interpretation of the CONTRACT DOCUMENTS.

28. LAND AND RIGHTS-OF-WAY

28.1 Prior to issuance of NOTICE TO PROCEED, the OWNER shall obtain all land and rights-of-way necessary for carrying out and for the completion of the WORK to be performed pursuant to the CONTRACT DOCUMENTS, unless otherwise mutually agreed.

28.2 The OWNER shall provide to the CONTRACTOR information which delineates and describes the lands owned and rights-of-way acquired.

28.3 The CONTRACTOR shall provide at his own expense and without liability to the OWNER any additional land and access thereto that the CONTRACTOR may desire for temporary construction facilities, or for storage of materials.

29. GUARANTY

29.1 The CONTRACTOR shall guarantee all materials and equipment furnished and WORK performed for a period of one (1) year from the date of SUBSTANTIAL COMPLETION. The CONTRACTOR warrants and guarantees for a period of one (1) year from the date of SUBSTANTIAL COMPLETION of the system that the completed system is free from all defects due to faulty materials or workmanship and the CONTRACTOR shall promptly make such corrections as may be necessary by reason of such defects including the repairs of any damage to other parts of the system resulting from such defects. The OWNER will give notice of observed defects with reasonable promptness. In the event that the CONTRACTOR should fail to make such repairs, adjustments, or other WORK that may be made necessary by such defects, the OWNER may do so and charge the CONTRACTOR the cost thereby incurred. The Performance BOND shall remain in full force and effect through the guarantee period.

30. ARBITRATION

30.1 All claims, disputes and other matters in question arising out of, or relating to, the CONTRACT DOCUMENTS or the breach thereof, except for claims which have been waived by the making and acceptance of final payment as provided by Section 20, shall be decided by arbitration in accordance with the Construction Industry Arbitration Rules of the American Arbitration Association. This agreement to arbitrate shall be specifically enforceable under the prevailing arbitration law. The award rendered by the arbitrators shall be final, and judgment may be entered upon it in any court having jurisdiction thereof.

30.2 Notice of the demand for arbitration shall be filed in writing with the other party to the CONTRACT DOCUMENTS and with the American Arbitration Association, and a copy shall be filed with the ENGINEER. Demand for arbitration shall in no event be made on any claim, dispute or other matter in question which would be barred by the applicable statute of limitations.

30.3 The CONTRACTOR will carry on the WORK and maintain the progress schedule during any arbitration proceedings, unless otherwise mutually agreed in writing.

31. TAXES

31.1 The CONTRACTOR will pay all sales, consumer, use and other similar taxes required by the law of the place where the WORK is performed.

Bibliography

Acret, J. *California Construction Law Manual,* 3rd ed. Colorado Springs, CO: Shepard's/McGraw-Hill, 1982.

The American Society of Civil Engineers. *Proceedings of the Specialty Conference on Construction Risks and Liability Sharing,* Scottsdale, AZ, January 24–26, 1979.

Barrie, D.S. *Directions in Managing Construction.* New York: John Wiley & Sons, 1981.

Bush, V.G. *Safety in the Construction Industry: OSHA.* Va. and Englewood Cliffs, NJ: Reston Publishing Co. & Prentice Hall, 1975.

California State Water Resources Control Board. *Prevention, Management and Resolution of Construction Contractor Claims on Clean Water Grant Projects, Vol. I—Discussion.* Division of Water Quality, Sacramento, CA, June 1980.

Clough, R.H. *Construction Contracting,* 3rd ed. New York: John Wiley & Sons, 1975.

BIBLIOGRAPHY

The Construction Industry Affairs Committee of Chicago: *CIAC Recommendations.* 228 North LaSalle Street, Chicago, IL, 60601. (Joint committee of American Institute of Architects, American Subcontractors Association, Builders Association, Consulting Engineers Council of Illinois, Construction Specifications Institute, and Mechanical Specialty Contractors Association.)

The Construction Specifications Institute. Washington, DC, 1985.

Department of Defense. *Value Engineering.* Handbook 5010.8-H. Washington, DC: Superintendent of Documents, U.S. Government Printing Office, Sept. 12, 1968.

Federal Acquisitions Institute. *Principles of Government Contract Law.* Washington, DC: The Office of Management and Budget and the Office of Federal Procurement Policy, September 1979.

Fisk, E.R. *Construction Engineers Form Book.* New York: John Wiley & Sons, 1981.

Fisk, E.R. *Construction Project Administration,* 3rd ed. New York: John Wiley & Sons (to be published in 1988).

Fisk, E.R. "Management Systems for Claims Protection." Technical paper presented at the ASCE Construction Division Committee on Professional Construction Management Specialty Conference on Engineering and Construction Projects—The Emerging Management Roles, New Orleans, LA, March 17–19, 1982.

Fisk, E.R. "Risk Management and Liability Sharing." Technical paper presented at the Annual State Conference of the Arizona Water & Pollution Control Association (AWPCA), Lake Havasu, AZ, May 1–3, 1985.

Irwin, W.J., P.E. "The Return of Eichleay: Is It Here to Stay, Part II" and "Denial of Time Extension is a Constructive Acceleration." *Construction Claims Monthly* (July 1984).*

Lambert, J.D., and White, L. *Handbook of Modern Construction Law.* Englewood Cliffs, NJ: Prentice-Hall, 1982.

Levin, P. *Claims and Changes.* Silver Spring, MD: WPL Associates, Inc., 1978.

---

* Abstracts of the above article appearing in Chapter 7 were reprinted with permission from *Construction Claims Monthly,* 951 Pershing Drive, Silver Spring, MD 20910, B.M. Jervis, Esq., Ed., (301) 587-6300.

BIBLIOGRAPHY

McDonald, P.R., Esq. "Recovery of Home Office Overhead—A Different Point of View." *Construction Claims Monthly* (December 1983).*

Teets, R.L. *Profitable Management for the Subcontractor.* New York: McGraw-Hill, 1976.

Thompson, L.J., and Tannenbaum, R.J. "Survey of Construction-Related Trench Cave-Ins." *Journal of the Construction Division,* Proceedings of the American Society of Civil Engineers, Vol. 103 (September 1977), p. 511.

U.S. Army Management Engineering Training Agency. *Principles and Applications of Value Engineering.* Department of Defense Joint Course Book, Vol. I. Rock Island, IL.

U.S. Department of Transportation. *Work Zone Traffic Control Standards and Guidelines.* Part VI of *Manual on Uniform Traffic Control Devices for Streets and Highways* (ANSI D6.1-1978). Washington, DC: Federal Highway Administration.

Watt, R.G., Esq., and Romm, D.C., Esq. "Recent Decisions Affecting Recovery of Home Office Overhead." *Construction Claims Monthly* (August 1983).*

---

* Abstracts of the above articles appearing in Chapter 7 were reprinted with permission from *Construction Claims Monthly,* 951 Pershing Drive, Silver Spring, MD 20910, B.M. Jervis, Esq., Ed., (301) 587-6300.

INDEX

Note: **Boldfaced** page numbers indicate principal discussion of the subject.

AAA, *see* American Arbitration Association, (AAA)
AASHTO, *see* American Association of State Highway and Transportation Officials (AASHTO), 58
AASHTO Standard Format for Highway Construction, 58
ABC, *see* Associated Builders and Contractors (ABC)
Abnormal weather, 145, 146
Absorption rate, 235, 236
    comparative, 234
    computation, 235
Acceleration, 36
    constructive, 36
    or suspension of the work, 36
    of the work, 36, 192, **243**
Acceptance periods, 261
Acceptance of the work, 269
Access:
    to project site, impact on bidding, 214
    to special standards, **83**

Accidents, 34
    records, **125**
    report forms, 179
Accounting systems, 180
ACI, *see* American Concrete Institute (ACI)
Acts of God, 32
Addenda, 168
Addenda to specifications, 71
Addendum to specifications, 73
Additional insureds, 157
Additional medical insurance, 282
Adjustment, equitable, 197
Administration, 170
    of contracts, 13
Administrative procedures, 13, **235**
Affirmative action, 109
AGC, *see* Associated General Contractors of America
Agency, limits of authority, 93
Agency delays, 174
Agency's right to correct, 226
Agency's right to remedy, 226

INDEX

Agency superior knowledge, 148
Agenda for preconstruction conference, **161**
Agreement, 168
AIA, *see* American Institute of Architects (AIA)
Air pollution control, 191
AISC, *see* American Institute of Steel Construction (AISC)
AISI, *see* American Iron and Steel Institute (AISI)
AITC, *see* American Institute of Timber Construction (AITC)
Allocation:
 of contract price, 149
 of risk, **29**, 39
All-risk builder's insurance, 285
Also-named insureds, 157
Altered bonds, 139
American Arbitration Association (AAA), **291**
 rules, 291
American Association of State Highway and Transportation Officials (AASHTO), 57, 58, 82
American Concrete Institute (ACI), **82**
American Institute of Architects (AIA), 49, 52, 65
American Institute of Steel Construction (AISC), 82
American Institute of Timber Construction (AITC), 82
American Insurance Association, 85, 154
American Iron and Steel Institute (AISI), 82
American Mutual Insurance Alliance, 154
American National Standards Institute (ANSI), 80, 128
American Plywood Association (APA), 82
American Public Health Association, 85
American Public Works Association (APWA), 82
American Society of Civil Engineers (ASCE), 46, 53, 129
American Society of Heating, Refrigerating, and Air-Conditioning Engineers (ASHRAE), 83
American Society of Mechanical Engineers (ASME), 83, 85
American Society for Testing and Materials (ASTM), 76, **79**, 87
American Water Works Association (AWWA), **80**
American Welding Society (AWS), 83
American Wood Preservers Association (AWPA), 83

American Wood Preservers Institute (AWPI), 83
Amount of liquidated damages, 274
Analysis phase, valve engineering, **218**
ANSI, *see* American National Standards Institute (ANSI)
Antibrokerage requirement, 136
Anti-Racketeering Act of 1946, 109
APA, *see* American Plywood Association (APA)
Approvals
 shop drawings, 173
 submittals, 173
APWA, *see* American Public Works Association (APWA)
Arbitration, 179, 185, 225, 248
 evidence, 185
 vs. litigation, 248
Architect-engineer, 32, 87, 147, 182, 210
 delays, 174
 interpretation of documents, 240
 judge of equality, 15
 negligence, 34
 product substitutions, 15
 rejection of non-conforming work, 207
 stopping of the work, **195**
 suspension of the work, **195**
Armed Services Board of Contract Appeals (ASBCA), 145
As-built drawings, 87, 265
ASCE, *see* American Society of Civil Engineers
ASHRAE, *see* American Society of Heating, Refrigerating, and Air-Conditioning Engineers (ASHRAE)
ASME, *see* American Society of Mechanical Engineers (ASME)
Associated Builders and Contractors (ABC), 114
Associated General Contractors of America, 27, 52, 66, 111, 114, 129, 154, 277
Assumption of risk, 22, 24
ASTM, *see* American Society for Testing and Materials (ASTM)
ASTM Standards, 169
Attorneys, 228
Audits, governmental, 178
Available hoisting facilities, 162
Award, 152
 and contract, **152**
Award procedure, 152
AWPA, *see* American Wood Preservers Association (AWPA)
AWPI, *see* American Wood Preservers Institute (AWPI)

INDEX

AWS, *see* American Welding Society (AWS)
AWWA, *see* American Water Works Association (AWWA)

Bar charts, 182, 193
Base costs, 233
Basic Building Code, 85
Beneficial use, 247, **265**
  definition, 266
Bibliography, 333
Bid, 14, 168
  bond, 12, 14, 18, 143, 168, 279
  bond form, 139
  certificate of mailing, 17
  contingencies, 39
  errors, 144
    relief from, 19
  filing, 105
  forfeiture of bond, 19
  informal, 47
  mailing, 17
  mistake, 15, 19
  nonresponsive, 2, 47, 153
  not correctable, 1
  opening, 143
  postmarks, 143
  prices, negotiations, 15, **146**
  publicly advertised, 14
  qualified, 2
  responsive, 14
  security, *see* Bid, bond
  sheets, 168
  shopping:
    effect of, 103
    or peddling, 17, **102**, 105, 135
  unit price, 7
  withdrawal of, 19
Bidder, 12, 163
  log at opening of bids, 145
  lowest responsible, 163
  lowest responsive, 163
  rejection, 12
  responsible, 12
  responsive, 12
Bidding, 134
  access to project site, 136, 214
  addenda to specifications, 73
  advertising, 12, 142
  award, 135
  barring from, 19

bid bonds, 143
bid shopping or peddling, 102
bid submittal, 142
certificate of mailing, 17
climate, 214
competitive process, 103
considerations, 214
contract documents, 134
control of bid shopping, 104
correctable errors, 139
disclaimer(s), **137**
  of information, 137
  of subsurface conditions, 138
drill logs, 138
effect(s):
  of addenda, 73
  of bid shopping, 103
  of weather on price, **145**
errors:
  excusable, 15, 144
  of fact, 144
  of judgment, 15, 144
geology, 214
informal bids, 139
integrity, 13
labor supply, 214
local services, 214
lowest responsive, responsible bidder, 139
lump sum, 140
materials and equipment, 214
mistake, 15, 19
nonresponsive bids, 139
opening of bids, 14, 143, **145**
payment of bonuses, 214
prebid inspections and conferences, **136**
preparing and submitting bid, **138**
price negotiations, 146
process, 13
public advertisement, 14
qualifying bid, **138**
rejection:
  of all bids, 139
  of bid, 139
site information, 138
soils reports, 138
source of materials, 214
statutory relief, 19
storage facilities, 214
substitute construction equipment, 214
system, integrity, 13

INDEX

topography, 214
unexcusable errors, 15
unit price, 140
Binding arbitration, 248
Board of Contract Appeals (BCA), 148
Boiler, machinery, and power plant insurance, 286
Boilerplate, 46, 48
Bond, labor and material, *see* Payment, bond
Bonding, 17
  capacity, 18
  considerations, 280
  factors, 280
  requirements, **17**, 157
Bonds, 12, 14, 178, 152, 205
  amount, 18
  bid, 12, 14, 18, 279
  checklist, 154
  forced placement, 281
  guarantee, 265
  labor and material payment, 280
  maintenance, 265, 280
  payment, 12, 14, 18, 157, 206
  performance, 12, 14, 18, 38, 157, **204**, 279
  proposal, 279
  release of Lien, 280
  subcontractor, **250**
  subcontractor performance, 250
  submittals, 153
  types, 279
  vs. insurance, 280
Brand-name product, 13
Breach of contract, 247
Broad form comprehensive liability policy, 282
Broker, contractors, 154
Building code:
  current edition, 85
  what and when adopted, 85
Building codes, regulations, ordinances, and permits, **84**
Building Officials and Code Administrators International (BOCA), 85
Building permits, 84
Burden fluctuation, 234, 236
  computation, 237
Burden of risk, 27
Business enterprises: minority, women, disadvantaged, 14
Business records, 177
Buy American, 14

Calculation:
  of bonding capacity, 18
  of overhead, 232
Calendar days, 194
California Department of Transportation (Caltrans), 60, 102
  special provisions format, **60**
  specifications, 60
Capability-related risks, **32**
Cardinal change, 246
Care, custody, and control, 289
Cash flow, 15
Cash flow records, 180
Cave-ins, 129
CBD, *see* Commerce Business Daily (CBD)
Certificate:
  completion, 263
  compliance, 265
  final payment, 264
  inspection, 265
  mailing, 17
  substantial completion, 268, 277
Certifications, 168
Certified payroll, 178
Change conditions, 38, 241
Change in construction method or sequence, 240
Change order(s), 14, 140, 162, 168, 173, **197**, 198, 254
  administration, **197**
  constructive, 173
  control sheet, 201
  delayed damages, 245
  evaluation, 201
  failure to agree on pricing, **245**
  form, 199
  log, 201, 202
  pricing, **245**
  request, 200, 201
  requirements, 198
  sheets, 203
  subcontracts, **201**
Changes:
  cardinal, 247
  constructive, 240
Checklist:
  bonds, **279**
  insurance, **281**
  surety and insurance bond, 279
Civil Rights Act of 1964, 106

INDEX

Claim(s), 3, 14, 42, 177, 179, 180, 207, **225**
  architect-engineer, 228
  burden of proof, 255
  construction funds, 275
  cost isolation data, 251
  delayed, 230
  disputes, **225**
  documentation, 250
    checklist, 255
    cost isolation data, 251
    to support, **254**
    types, 255
  evidence, **255**
  formal notice, 15
  importance of notice, 226
  issues, 229
  method of presentation, 255
  notice, 225, 226
  records, **179**, 227
  resolution:
    alternatives, 258
    methods, 257
  right to file, **227**
  subcontractor, 250
  support records, **180**
  waiver of, 15
  when to make, 227
  witness checklist, 256
Clauses, 28
  exculpatory, 7, **27**, 28, 38, 39, 40
  liquidated damages, 7
Cleanup of project site, 162, 264
Clerical error, 19
Climate, impact on bidding, 214
Code of Ethical Conduct of the Associated General Contractors of America, 104
Code of Federal Regulations (CFR), 95, 118
Collective bargaining, **110**
Commerce Business Daily (CBD), 17
Communications, 170, **190**, 214
Comparable quality and utility, 15, 147
Comparative absorption rate, 234
Comparing bids, 7, 141
Compensable delay claims, 230
Competitive bidding, 13, 103
Completion:
  substantial, **267**
  of the work, **260**, 264
Completion date, 197

Compliance:
  with equal employment opportunity requirement, 178
  with laws and regulations, **91**
  with special laws and regulations, **158**
Compliance certificates, 265
Component parts of specification, **46**, **47**
Comprehensive general liability insurance, 113, 282
Compromise, 16, 253
Comptroller general, 104
Computer cost reports, 180
Concrete Reinforcing Steel Institute (CRSI), 83
Conditional release, 205, 276
Conditional waiver, 276
Conditional waivers of lien, 265
Conditions:
  changed, 38, 241
  of contract, 50, 68
  differing site, 38, 42, 197
  giving rise to contract changes, 198
  latent, 242
  unforeseen, 22, 241
  unforeseen underground, **196**
  unknown physical, 197
  unknown subsurface, 242
Conference:
  construction coordination, 159
  preconstruction, 159, 172
Conflicts:
  arising from addenda to specifications, 73
  in plans and specifications, **44**, **246**
Conforming work, 195
Consent of sureties, 265
Consent of surety for final payment, 265
Considerations when bonding, 280
Construction:
  bar chart, 183
  claims, 40
  contracts, **9**
  coordination conference, 159
  costs, 214
  diaries, 182
  disputes:
    compromise, 253
    negotiation, 253
    resolution, **253**
  field office, 190
  funds, claims on, 275
  Industry Affairs Committee (CIAC), 277

INDEX

Industry Arbitration Rules (AAA), 291
law, purpose, 92
lighting, 189
method delays, 231
progress record, 183
progress report S-curve, 184
related risks, **25**
risk allocation, 30
risks, 22
safety, 41
  and health regulations, 118
Specifications Institute (CSI), 46, 52
Construction Specifications Institute (CSI), 46, 52
  classification numbers, 53
  division/section concept, 52, **54**
  format, divisions, 52
  16-division format, **52**
  specifications format, **52**
  three-part section format, 48, **54**, **57**
Constructive acceleration, 36
Constructive change order, 173
Constructive changes, 173, **240**
  causes, 240
  definitions, 240
Constructive suspension, 36
Contents of specifications, 46
Contract(s), **152**
  of adhesion, 146, 246, 253
  billings, 233
  breach, 247
  change(s), 198
    reasons, 198
  claims, litigation, 39
  completion date, 195
  construction, 9
  cost-plus, 7
  cost-plus-fixed-fee, 7
  cost-plus-percentage-of-cost, 7
  documents, 44, 46, 51, 64, 100, 134, **168**, 193, 260, 261, 270
    sample public works, 303
  drawings, 86
  incentive, 7
  law, 94
  limits, 206
  lump sum, **5**
  price, 7, **152**
  prime, 155
  procedure, 152

scope, 206
time, **194**
  and materials, 7
types of, **5**
unit price, **6**
Contractor(s), 3, **171**
  broker, 154
  capacity, 29
  default, 38, 269
  delay, causes, 231
  as insurer, 25, 28
  liability, 28
  protection, **27**
  public works, 3
  rights, **273**
  risks, **40**
  superintendent, 260
  team, 178
  value engineering, 41
  work, **260**
Contractual allocation of risk, 27
Contractual legal risks, **25**
Contractual occasion of risk, 37
Contractual relationship, 20
Contractual safety requirements, 124
Contractual shifting of risk, 26
Contractual status:
  subcontractors, **20**
  suppliers, **20**
Control(s):
  of bid shopping, **104**
  environmental, **190**
Control Sheet, Change Order, 201, 203
Coordination, **192**
  of subcontract and prime contract, 249
Copeland Act of 1934, **108**
Corps of Engineers (Department of the Army), 99, 101
Correctable bidding errors, 139
Correspondence, 180, 197
Cost(s):
  impact, 246
  isolation data, 227
  isolation documentation, 252
  isolation records, 245
  job, **180**
  mobilization/move-on, 206
  monitoring, **185**
  overrun, 189
  overruns, 140

· 341 ·

INDEX

Cost(s) (*Continued*)
  records, 179
  reporting, **194**
  segregation, 227
  sharing, 41
Cost-loaded CPM schedules, 194
Cost-plus contract, 7
Cost-plus-fixed-fee contract, 7
Cost-plus-percentage-of-cost contract, 7
Costs and expenses, segregating and tracking, 179
Courts, 225
CPM and cost reporting, 194
CPM network, 193
CPM schedules, 193
Critical path method (CPM), 192, 193, **194**, 239
  scheduling, 192, 239
CRSI, *see* Concrete Reinforcing Steel Institute (CRSI)

Daily construction diary, 186
Daily construction report, 180
Daily log, 181
Daily report, 180, 181, 185
Daily superintendent forms, 179
Damages, 42, 177, **249**
  for delay, **249**
  elements, 273
  liquidated, 42, 177, 195, 240
  measure, 273
Davis–Bacon Act, 95, **107**, 112
Deceleration, 239, 246
Decisions, architect and engineer, 15
Deductibles, 283
Default, 269
  contractor, 38
Default claims, 153
Defective plans and specifications, 240
Defective work, 34
Defects:
  design, **33**, 34
  specification, 34
Defend and hold harmless, 39
Defense, 39
Delay(s), 180, 192, 195, 196, 207
  causes, 231, **272**
  claims:
    compensable, 230
    excusable, 230
    nonexcusable, 230
  damages for, 249

damages and change orders, 246
  on the job, 196
  liquidated damages, 7, **271**
  memorandum of, 244
  owner-caused, 229
  submittals, 174
  unreasonable, 273
  weather, 145
Delivery schedule, 155
Demobilization, 264
Department of the Army, Corps of Engineers, 99
Department of Defense, 78
Department of Labor, 112
  wage rates, 14
Department of Transportation, 138
Department of Transportation Contract Appeals Board, 146
Design, substandard, 22
Design change delays, 231
Design defects, 34
Design failures, 34
Design professions, 33
Detailed cost records, 227
Determination of bond amount, 18
Development phase, value engineering, **219**
Deviations in shop drawings from plans and specifications, 174
Diary, 185, 186
Differing site conditions, 31, 38, 42, **195**, 197, 241
Direct cost, 195, 245
Disadvantaged business enterprises, 14
Disclaimer of bidding information, 137
Disclaimer provisions, 38
Disclaimer of subsurface conditions, 138
Disclaimers in bid solicitation, **137**
Disclosure of site information, 148
Disposal of hazardous waste, 191
Disputes, **42**, 179, **225**, 230, 248
  formal notice, 15
  notice, 226, 248
  procedural requirements, 248
  records, 179
  resolution, 247, 255
  time, 230
  waiver, 15
Documentation, **19**, 177
  by camera, 251
  claim **250**, 255
  contract, **168**, 255

INDEX

cost isolation, 252
daily report, 251
delays in the work, 251
differing site conditions, 251
disputed work, 251
extra work ordered, 251
field diary, 251
importance of, **177**
organizing support for claim, **254**
survey request, 188
system, 182
theory, 251
types, 255
use of, **177**
by video, 251
Documentation interpretation, 240
Drawings, **43,** 46, 47, 179
  as-built, 265
  conflicts with specifications, **44**
  contract, 86
  record, 265
  record/as-built, 87
  shop, 87
  standard, 86
  types in construction contract, **86**
Drill logs, 138
Dust abatement measures, 191

Early completion, 7
Earned funds, 206
Earthquake, 32
Economic disasters, **35**
Economic risks, **25**
Effect(s):
  of bid shopping, 103
  of weather on bid price, **145**
Eichleay Formula, 232, 234, 236, 238
EJCDC, *see* Engineer's Joint Contract Documents
  Committee (EJCDC)
Employee time records, **180**
Employer liability insurance, **113**
Endorsements, insurance selection, 282
Engineering economics, 210
Engineer's errors and omissions, 27
Engineers' estimate, 7
Engineer's Joint Contract Documents Committee
  (EJCDC), 27, 49, 65, 204
Environmental control, **190**
Environmental Protection Agency (EPA), 92
  Clean Water Grant Program, 92
Environmental risks, 37

Equal Employment Opportunity (EEO), 109, 178
Equal Employment Opportunity Commission
  (EEOC), 106
Equal Opportunity Employment Act of 1972,
  106
Equipment, 36
Equipment instruction, 265
Equipment testing, 261
Equitable adjustment, 197
Error(s):
  bidding, 15
  excusable, 15, 19
  of fact, 19, 144
  of judgment, 15, 19, 144
  and omissions, 27, 28
  unexcusable, 19
Esthetics, architect-engineer, 15
Estimate(s), 180
  engineers', 7
Ethnic minorities and women in construction,
  109
Evidence, 185
Excavation safety, *see* Trench and excavation
  safety
Exclusions, 289
Exculpatory clauses, 22, **27, 28,** 38, 39, 40
Excusable delay, 195, 230
Excusable error, 15, 19, 144
Executive Order, 11246, **107**
Exhaustion of administrative remedies, 225
Expert witness, 254
Explosives and blasting, 191
Extended home office overhead, 195
Extensions of time, 273
Extra work in subcontract, **201**
Eyewitness, 254

Fabricators, 171
Facilities, temporary, **189**
Factors in bonding, 280
Failure to make payments, 247
Fair labor standards, 108
Federal Acquisition Regulations (FAR), 41, 141,
  143, 144
Federal Highway Administration, 97, 128
Federally-Assisted Water and Sewer Project,
  sample documents, 303
Federal Register, 108
Federal wage rates, 12, 47
Federal Water Pollution Control Acts, Amendments of 1972, 101

· 343 ·

INDEX

Field inquiry delays, 231
Field offices, **190**
Field Order, 204
Field telephones, **190**
File memos, 179
Final acceptance, 269, 270, 276
Final contract price, 7
Final inspection, 264
   notice, 263
Final payment, 206, **269, 270**
   conditions, 276
   as release, 276
   as waiver, 270
   withholding from, **269**
Final progress payment, 206
Final punch list inspection, 264
Final submittals, 265
Final waivers of lien, 265
Financial failure, 35
Financial records, 178
Fire, 32
Float, scheduling, 240
Floater policies, 286
Flood, 32
Force account, 226
Forced placement of bonds and insurance, 281
Forecasts, 180
Forfeiture of bid bond, 19
Formalities of public contracts, 19
Formal notice: claims, protests, disputes, 15
Form of payment request, 161
Form of schedules, **193**
404 Permit, Corps of Engineers, 101
Fundamentals of value engineering, 211
Funding, **36**

General conditions, 27, 44, 48, 49, 50, 51, **64,** 65, **66,** 168, 171
   typical subjects, 64
General contractor(s), 34
   antibrokerage requirement, 136
   as "broker," 104
General contractors' responsibility, 172
General partnership, 8
General project status report, 187
General Provisions, 64
General requirements, 50
General Services Administration (GSA), 104, 148
   public buildings service, 102
Generating value engineering proposals, 215

Geological formation forecasts, 39
Geological information, 38
Geology, 31
   impact on bidding, 214
Government audit, 178
Government code claim, 226
Government standards, **76**
   federal specifications, **76**
   military specifications, 78
Groundwater, 31
Guarantee, commencement, 265
Guarantee bonds, 265
Guaranteed maximum price, 22
Guarantees and warranties, **158,** 265

Handling rejected work, **207**
High-risk construction, 26
High-risk underground work, 26
Hobbs Act, **109**
Home office overhead, **232,** 233
Hurricane, 32

ICBO, *see* International Conference of Building Officials
ICBO Plumbing Code, 85
ICBO Uniform Mechanical Code, 85
Identification of risk, 26
IES, *see* Illuminating Engineering Society (IES)
Illuminating Engineering Society (IES), 83
Impact cost, 195, 207, 228, 246
Impact of delay, 192
Importance of job site communications, 214
Impossibility of performance, 240
Improper construction delays, 231
Improper inspection, 240
Improper rejection of work, 240
Inadequate coordination delay, 231
Incentives, 7
Incorporation of prime contract, **155**
Incorrect plans and specifications, 28
Indemnification, 39
Indemnify, defend and hold harmless, 40
Indemnity language, 40
Indemnity provisions, 40, **156**
Indirect costs, 233
Inflation, 35
Informal bids, 47, 139
Information, pre-bid, 16
Information phase, value engineering, 215
In-place costs of operation, 211
Inspection, **261**

INDEX

Inspection, pre-final, 261
Inspection certificates, 265
Inspection delay, 231
Inspector, 171
Instructional requirements, 261
Instruction to Bidders, 47, 139, 168
Insurance, **152**, **161**, **281–289**
   accounts receivable, **288**
   all-risk builder's standard insurance, 284
   automatic builder's risk, 285
   automobile, 283
   automobile medical payments, 284
   automobile physical damage, 284
   boiler, machinery, and power plant, 286
   and bonds, **157**
   broker, 157
   business interruption, 288
   care, custody and control, 289
   checklist, **281**
   completed operations, 283
   completed value, 284
   comprehensive general liability, 113, 282
   contractor's equipment floater, 286
   contractual or assumed liability, 283
   coverage, 152
   deductibles, 283
   elevator, 23
   employer liability, **113**
   endorsements, 282
   exclusions, 289
   fire and explosion legal liability, 287
   floater policies, 286
   forced placement, 281
   independent contractors, 283
   installation floater, 286
   liability forms, 281
   miscellaneous, 287
   multi-peril crime, 287
   multiple-peril builder's risk insurance, 285
   nonstandard, 285
   ordinary builder's risk, 285
   perils, 285
   personal injury, 283
   policies, 12, 178
   premises operations, 282
   products liability, 283
   property damage, 12
   property forms, 284
   public liability, 12
   reporting basis, 284
   right of subrogation, 288
   standard builders' risk, 284
   submittals, 153
   subrogation, 288
   transportation floater, 286
   umbrella excess liability, 283
   valuable papers destruction, 288
   water leak and liability, 287
   workers' compensation, **113**, 281
Insured Perils, 285
Interference delays, 231
International Association of Plumbing and Mechanical Officials (IAPMO), 85
International Conference of Building Officials (ICBO), 84, 85
Interpretation of documents, architect-engineer, 240
Invoices, 180

Job acceleration, 179, 239
Job changes, **204**
Job costs, **180**
Job costs accounting system, 180
Job deceleration, 239
Job interference, 239
Job progress scheduling, 161
Job security during nonworking hours, 162
Job site communications, importance, 214
Job site security during nonworking hours, 161
Joint checks, 206
Joint and several liability, 8
Joint ventures, **8**
Journal of the Construction Division, ASCE, 129
Judge of equality, architect-engineer, 15
Jurisdictional levels, 91

Keying schedule, 265
Key provisions for subcontractors, 156

Labor, 36
   agreement, *see* Labor, contract
   agreement risks, 41
   code, 130
   considerations, 111
   contract, 111
   laws, **105**
   and material payment bonds, 206, 280
   materials, and equipment, **36**
   nonunion, 110
   relations, **110**
   releases, **205**

INDEX

Labor (*Continued*)
  secretary, 121
  supply impact on bidding, 214
  union, 110
lack of right-of-way delay, 231
Landrum–Griffin Act, **106**
late approval delays, 231
late bid, 17
  mailing, 17
late completion, 206
latent conditions, 196, 197, 242
late payments, 247
late procurement delay, 231
late response delays, 231
laws, 95
  affecting subcontracts, **101**
  Anti-Racketeering Act of 1946, 109
  Civil Rights Act of 1964, 106
  controlling traffic during construction, 97
  Copeland Act of 1934, **108**
  Davis–Bacon Act, 95, **107**, 112
  Equal Employment Opportunity Act of 1972, **106**
  Executive Order 11246, **107**
  Federal Water Pollution Control Acts, Amendments of 1972, 101
  governing execution of the Work, 94
  Landrum–Griffin Act, **106**
  licensing, 94
  National Labor Relations Act, 110
  protecting subcontractors, 14
  protective, 13
  public contracts, 9
  Public Works Employment Act of 1977, 109
  relating to settlement of differences and disputes, 94
  Rivers and Harbors Act of 1899, 99, 101
  Sherman Anti-Trust Act, 106
  subcontractor protection, 14
  Title VII—Equal Employment Opportunity, 106
  Williams–Steiger Occupational Safety and Health Act of 1970, 118
Lawyers, 228, 229
Legal review, 228, 229
Levels of risk, 24
Liability, joint and several, 8
Liability forms of insurance, 281
Liability sharing, 22, **24**
Licensing laws, **11**, 94
Liens and lien laws, 275

Lien waivers, 265, **276**
Limitations of authority of public agency, **93**
Limits:
  bonding capacity, 18
  of insurance coverage, 152
Liquidated damages, 7, 42, 177, 195, 240, 269, 271, **273**, 274
  after acceleration, 274
  amount, 274
  clauses, 7
    in subcontracts, **274**
  late completion, 206
  unenforceable, 274
List of equipment suppliers, 168
List of subcontractors, 13, 17, 168
Litigation, 179, 185, 227, 248, 254
  construction claims, 39
  evidence, 185
  eye witness, 254
Local services impact on bidding, 214
Log, submittal, 174
Logic for risk decisions, 23
Longshoreman's insurance, 282
Losses, mitigating, **29**
Lost time accident report, 126
Low bidder, 12, 19, 144
Lowest responsive, responsible bidder, 12, 139, 163
Lump-sum bid, 149
Lump-sum contracts, **5**, 140, 141

Mailing bids, 17
Maintenance, 211, 215
Maintenance bonds, 265, 280
Maintenance of stock items, 265
Managerial competence, **34**
Managing risk, 24
Manuals, operating, 265
Manual on Uniform Traffic Control Devices, 128
Manufacturer's certificates, 207
Material(s), 36, **287**
  bond, 206
  certificates, 207
  delivered but not yet installed, 206
  delivery of, 206
  installed, 206
  receipts, 180
  releases, **205**
  standards, 75–83
  suppliers, 12, 20, 176
MBE, *see* Minority Business Enterprise (MBE)

INDEX

Measurement of quantities, 51, 149
Mechanic's lien laws, 275
Memorandum of delay, 244
Memos, 179
Method of construction, 169, **207**
Military specifications (Department of Defense), 78
Mil-specs, *see* Military specifications (Department of Defense)
Minimizing risks, **29**
Minimum-wage rates, 47
Minority Business Enterprise Provisions, 178
Minority Business Enterprise(s) (MBE), 14, 109, 179
Miscellaneous insurance, 287
Mistakes in bidding, 15, 19
Mitigating losses, **29**
Mobilization, 71, 206
Monthly progress payments, 51, 194
Monthly report of project time and cost status, 187
Move-on costs, 206
Multiple jurisdictions over construction project, 91
Multiple-peril builder's risk insurance, 285
Multiple prime contractors, 20

National Aeronautics and Space Administration (NASA), 53
National Apprenticeship Act of 1937, 109
National Building Code, 84
National Electrical Code (NEC), 85
National Electrical Manufacturers Association (NEMA), 83
National Fire Protection Association (NFPA), 83
National Historic Preservation Act, 191
National Labor Relations Act, 110
National Labor Relations Act of 1935, 106
National Plumbing Code, 85
National Society of Professional Engineers (NSPE), 52
Nationwide strikes, 35
Nature of risks, 26
Negligence, architect-engineer, 34
Negotiation, 14, 253
  of bid prices, **146**
NEMA, *see* National Electrical Manufacturers Association (NEMA)
NFPA, *see* National Fire Protection Association (NFPA)
Noise abatement, 191

Noncollusion affidavits, 168
Nonconforming work, 195, 207
Nondisclosure by agency, 247
Nonexcusable delay claims, 230
Nonexcusable delays, 195
Nongovernment standards, 79
Nonpayment, 35
Nonresponsive bids, 2, 47, 139, 153
Nonstandard insurance, 285
Nonunion labor, 110
Normal weather, 32, 145
Notice
  abatement provisions, 191
  acceptance, 264
  change order, 205
  claims, 225, 226
  completion, 264
  of dispute, 226
  importance of, 226
  inviting bids, **2**, 12, 145, 168
  potential claims, 226
  proceed, **163**
  requirement for, 205
  stop, 20
NSPE, *see* National Society of Professional Engineers (NSPE)

Objection to surety, 143
Occupational Safety and Health Administration (OSHA), 87, 95, 118
Office overhead, 240
Offshore work, 100
One-to-one concept, **170**
On-site relations, **182**
On-site safety, responsibility, **116**
On-site transactions, **182**
Open bidding, 14
Opening of bids, **145**
Open-shop contractors, 112
Operating manuals, 265
Oral agreements, 254
Ordinances, public contracts, 9
Or-equal provisions, 13, 14, 15, 16, 147, 172, 208
Or-equal submittal data, 147
Original estimates, 180
OSHA, Occupational Safety and Health Administration (OSHA)
OSHA Construction Safety and Health Regulations, 131
  categories of control, 119

OSHA Construction Safety and Health Regulations (*Continued*)
  code compliance, 121
  and the contractor, **117**
  enforced by labor secretary, 121
  fines, 124
  inspection procedure, 121
  records requirements, 121
  reporting requirements, 125
  safety and health standards, 118
  violations, criminal nature, 124
Overhead, 195, **232**, 234, 236, 238, 245
  calculation of, 232
  differential, 238
  extended, 195
  home office, **23**, 195, 232, 233, 245
  home office extended, 195
  over-absorption of fixed overhead, 234
  and profit, 7
  unabsorbed home office, 195, 232, 238, 245
Overruns, 195
Owner-caused delay, 146, 229, 231
Owner delay, causes, 231
Owner disclosure of site information, **148**
Owner nondisclosure, 240

Paper trail, 19
Paperwork, 3
Partial utilization, **265**
Partnership debts, 8
Partnerships and joint ventures, **8**
Passing the risk, 38
Payment, **206**
  bond, 12, 14, 18, 157, 168, **206**
  final, 206, **269**
  measurement of quantities, 51
  monthly progress, 51, 194
  partial, 51
  payroll reports, 161
  progress, 51, 206
  recommendation for, 51
  requests, form of, 161
  schedule of, 197
  schedule of values, 149
Payrolls, 180
Performance:
  bonds, 12, 14, 18, 38, **154**, 157, 168, 169, **204**, 234, 279
  risk, 25
  specifications, **168**

standards, 75
time for, **154**
Perils, 285
Permit requirements, 153
Permits, 96, 99, **152**, 153, 161
  for activities in navigable waterways, 99
Philadelphia Plan, the, **107**
Philosophy of value, **213**
Phone memorandum, 179
Physical risks, **25**
Plans, defective, 240
Plans and specifications, 197, 240
  conflicts, 246
Political risks, 25
Political and societal risk, **36**
Postmarks on bids, 143
Potential claim or dispute, 225
  notice, 225
Pre-approved subcontractors, 104
Prebid conferences, **136**
Prebid information, 16
Prebid site inspections, **136**, 137, 148
Preconstruction conference, **158**, 161, 162, 163, 170, 172
  agenda, 160, **161**
  available hoisting facilities, 162
  benefits, 159
  change orders, 162
  job progress scheduling, 161
  payroll reports, 161
  progress payments, 161
  punch list procedures, 163
  time, 160
Precontract exploration, 31
Prefinal inspection, 261
Prejob labor considerations, **111**
Premium time, 243
Preparation:
  of claim file, 227
  of Punch List, 261
  and submittal of bids, **138**
Prescriptive specifications, **168**
  vs. performance specifications, **168**
Presentation and follow-up phase, value engineering, **220**
Prevailing wage rates, 12
Price:
  bid, 152
  comparing bids, 7
  contract, 152
  guaranteed maximum, 22

INDEX

Pricing, change orders, **245**
Prime contract, 39, 155
   incorporation of, **155**
Prime contractor(s), 20, 34
   multiple, 20
Principal laws governing public works construction, **94**
Privity of contract, 20
Problems, scheduling, 192
Procedure, contract, 152
Product(s):
   "brand name," 13
   "or equal," 13
   substitution of, **15**, 16, **147**, **208**
Production bonuses, 214
Product substitutions:
   architect-engineer, 15
   procedure, 147
Profit, 7
Progress charts, 182
   conditions, 276
   during construction, 206
   as release, 276
Progress record, 183
Progress reporting, **180,** 182
Project:
   closeout, **259**
   communication, 170
   drawings, 168
   manager, 170
   Manual (CSI), 46, 74
   multi-agency control, 86
   overruns, 195
   signs, **191**
   site, cleanup, 264
   specifications:
      CSI format, **62**
      vs. special provisions, **62**
   subject to control by more than one agency, **86**
Property damage insurance, 12
Property form of insurance, 284
Proposal, 168
Proposal bond, 279
Protection:
   of public, 3, 9, **128**
   for subcontractors, 14, 17
Protective legislation for subcontractors, 104
Protest:
   formal notice, 15
   waiver, 15

Provisions for temporary facilities, 70
Public advertising, 17
   of a project, **17**, 142
Public agency, limits of authority, 93
Public agency representative, 171
Public Buildings Service (GSA), 102
Public contracts:
   formality, 19
   laws, ordinances, regulations, 9, 13
   requirement for, 12
Public disorder and war, 37
Public funds, 9
Public interest, public works contracts, 9, 12
Public liability insurance, 12
Public protection of, 3
Public risks, **25**
Public Works Employment Act of 1977, 109
Punch list, 206, **260**, 270, 271
   completed, 206
   cost completion, 270
   form, 262
   inspection, final, **264**
   items, 206
   preparation, **261**
   procedures, 163, **260**
   uncompleted items, 206, **277**
Purchase orders, 180

Qualified bid, 2
Qualifying of bids, **138**
Quantities of work, 32
Quantity takeoffs, 180
Quantity variation delays, 231

Readvertising the job, 139
Reallocation of risk, 27
   to subcontractors, 39
Rebidding, 19
Recommendation for payment, 51
Record drawings, 87, 265
Recordkeeping, **158**, **178**, 179, 254
Recordkeeping, safety, 121
Records, 177, **178**
   claims and disputes, **179**
   claims support, **180**
   contract, **178**
   cost isolation, 245
   employee time, **180**
   financial, 178
   required by contract, **179**
   required by law, 178

INDEX

Recovery of damages, 273
Red tape, 13
Regulations, 37
  compliance with, 158
  public contracts, 9
Rejection:
  of bids, 139
  of nonconforming work, 207
Relations, on site, 182
Relative importance of different contract documents, 51
Release(s)
  conditional, 205, 276
  documents, 205
  labor, **205**
  lien bonds, 280
  material, **205**
  retainage, 206, **271**
  subcontractor, 276
  unconditional, 205, 276
Relief from bid errors, **19**, 144
Report, daily construction, 180
Reporting, progress, **180**
Request(s):
  for information, 179
  for services, **189**
  for survey, 188
Resident engineer, 171
Resolution of construction disputes, **253**
Responsibility for on-site safety, **116**
Responsible bidder, 12, **163**
Responsive bids, 14, **139**
Retainage, 15, 206, 264, 269, 277
  from earned funds, 206
  payment, 264
  release of, 206
Right to file claims, 227
Right of subrogation, 288
Right-to-work legislation, 110
Risk(s), 3, 24
  acceleration or suspension of the Work, 36
  accident, 34
  acts of God, 32
  allocation, **29**, 30, 157
  assumption, 22, 24
  burden of, 27
  capability-related, **32**
  contractual allocation, 27, 37
  contractual and legal, **25**
  decisions, 23
  defective design, 33
  defective work of construction, 34
  definition, 24
  disputes, 42
  dumping, 38
  economic, 25
  economic disasters, 35
  environmental, 37
  financial failure, 35
  funding, 36
  incorrect plans and specification, 28
  inflation, 35
  labor, materials, and equipment, 36
  levels of, 24
  and liability sharing, 24
  management, 22
  manager, 24
  managerial competence, 34
  minimizing, 29
  nonpayment, 35
  performance, 25
  physical, 25
  political and societal, 25, 36
  public disorder and war, 37
  quantity variations, 32
  reallocation, 27
    to subcontractors, 39
  regulations, 37
  reserved to contractor, 40
  sharing precepts, 24, 37
  shift of, 26, 27
  site access, 29
  subcontractor failure, 34
  subsurface conditions, 31
  transfer, 22
  types of, 29
  unforeseen conditions of, 35
  union problems, 37
  violation of laws, 28
  weather, 32
Rivers and Harbors Act of 1899, 99, 101
Rubbish control, 191
Rules, Construction Industry Arbitration (AAA), 291
Rules for diaries, 185
Rules of the game, 91

Safe subcontractors, 117
Safety:
  construction, 41
  contractual requirements, 124
  control, 29

INDEX

meetings, 119
plans, **119**
practices, 215
primary liability, **116**
program, 29, 119
protection of the public, **128**
record keeping, **121**
responsibility for, **116**
subcontractors, **117**
trench and excavation, **129**, 130
violation procedures, **121**
workers' compensation insurance rates, 113
Sanitary facilities, 190
Sample documents, public works, 303
Samples, submittals of, **172**
Sample submittal data, 161
Schedule(s), 180
  approvals, 239
  change delays, 231
  cost-loaded CPM, 194
  CPM, 193
  delivery, 155
  form of, **193**
  impacts, 239
  of payments, 197
  submittals, **193**
  of values, 6, 149, 164
Scheduling, 192
  changes, **239**
  coordination, **192**
  CPM, 192, 239
  float, 240
  techniques, 239
  trends, 182
Scope of insurance coverage, 152
S-curve, 182, 184
Security of bid, 19
Segregating costs and expenses, 179
Selection of subcontractors, **135**
Services, requests for, **189**
Settlement, 227
Sheeting, shoring, and bracing, 130
Sherman Anti-Trust Act, 106
Shifting of risk, **26**, **27**
Shop drawing(s), 87, 161, 172, 173
  approval, 87
  processing, 172
  review, 87
  and sample submittal data, 161
  submittals and proposals, 87, 172, 173
Signs, project, **191**

Site access delay, 231
Site conditions, differing, 180, **195**, **241**
Site information, obligation to disclose, 138, 148
Soils reports, 31, 138, 148
Sole negligence, 40
Source for materials impact on bidding, 214
Southern Building Code Congress (SBCC), 85
Southern Standard Building Code, 85
Special bonds, 265
Special Conditions, 67
Specialized subcontracting, 215
Special laws, compliance with, **158**
Special material and product standards, 75
Special provisions, 57, 58, **62**, **63**, 75
Specification(s), 44, 46, **168**, 197
  addenda, 71
  boilerplate, 46, 48
  California Department of Transportation, 60
  Caltrans format, 60
  contents, **46**
  contract documents, 64
  controlling contract documents, 46
  controlling criteria, **44**
  CSI Project Manual, **62**, 74
  CSI 16-division/section concept, **52**, **54**
  defective, 240
  defects, 34
  definition, **44**
  and drawings, 43
  federal government standards, 76
  format, 48
  General Conditions, 48, **64**
    limitations, 65–67
  General Provisions, **64**
  general subjects, 49
  limitations of Standard General Conditions, 65–67
  nonstandard formats, **61**
  performance standards, 75
  Project vs. Special Provisions, **62**
  provisions for temporary facilities, 70
  special material and product standards, 75
  Special Provisions, 75
  special standards, 83
  specifications, performance and prescriptive, **168**
  specifications writer, 46
  standard, 73
    limitations, 73–75

INDEX

Specification(s) (*Continued*)
  standard General Conditions limitations, 65–67
  standard provisions, 58
  state highway department formats, 54
  substandard, 22
  supplemental, 58, **63**
  supplemental provisions, 75
  Supplementary General Conditions, 67
  technical provisions, 46, **68**
  typical CSI format technical provisions, 68
  Uniform Building Code (UBC), 79
Speculation Phase, value engineering, **218**
Spill bank, 130
SSPC, *see* Steel Structures Painting Council (SSPC)
Standard builders' risk insurance, 284
Standard drawings, 86, 168
Standard General Conditions, limitations, 65–67
Standardized forms, 179
Standardized subcontracts, 154
Standard provisions, 58
Standards, ASTM, 169
Standards, materials and products, 75–83
Standard specifications, 46, **73**
  limitations, 73–75
State highway department formats, 54
State-licensing of contractors, 114
State wage rates, 47
Statutory claims procedures, 226
Statutory relief for bidder, 19
Steel Structures Painting Council (SSPC), 83
Stop Notice, 20, 275
Stop Notice Release Bond, 271
Stop payment orders, 275
Stopping of work by architect-engineer, **195**
Strict compliance with bidding requirements, 139
Subcontract(s), **101**, **155**
  award by agency, 105
  change orders, **201**
  consistency, 248
  coordination, 249
  damages for delay, **249**
  extra work, **201**
  key provisions, **156**
  liquidated damages, **274**
  standardized, 154
  unforeseen site conditions, **250**
Subcontracting, 154

Subcontractor(s), 12, 20, **101**, **154**, 156, **171**, 180
  agency approval, 104
  agreement, risks, 41
  arbitration, 248
  bonds, **250**
  contractual status, **20**
  delays, 231
  dispute clauses, **247**
  failure, 34
  identification, 14
  involvement with safety, **117**
  lien releases, **276**
  listing, 13, 17, 20
  litigation, 248
  Performance Bond, 250
  pre-approving, 104
  privity of contract, 20
  problems, 180
  protection of, 17
  protective legislation, 104
  relations, **247**
  safety, **117**
  selection, 14
  statutory protection, 20
  work, **260**
Submittal(s), **172**, 179
  approval, 173
  bonds, 153
  control sheet, 174, 176
  final, **265**
  insurance, 153
  log, 174, 175
  process, 173
  samples, **172**
  schedules, **193**
  shop drawing, 87
Subrogation, 288
Substandard specifications and design, 22
Substantial completion, 264, **267**
Substantial performance, 267
Substitution of products, **15**, **16**, **208**
Subsurface conditions, **31**, 196, 197
Superintendent, 170, 260
Superior knowledge, 148
Supervision of the Work, 170
Supplemental provisions, 75
Supplemental specifications, 57, 58, 63, 74
Supplementary General Conditions, 27, 50, 51, 67, 168
  typical contents, 68

· 352 ·

INDEX

Supplier certificates, 207
Suppliers, privity of contract, 20
Suppliers' contractual status, **20**
Surety(s), 143
   Association of America (SAA), 154
   bond checklist, **279**
   bonds, **279**
   consent, 265
   selection, 281
Survey request, 188
Suspension:
   constructive, 36
   of the Work, 36, 192, 195, **245**
   of the Work by architect-engineer, **195**
Systems, paperwork, 3

Taft–Hartley Act, 106
Target date, 195
Taxpayer, 13
Tax-rate changes, 35
Technical provisions:
   of specifications, 46, **68**
   typical subjects under CSI format, 68
Technical specifications, 168
Telephone messages, 179
Temporary facilities and controls, 161, **189**
Temporary pipelines, 190
Temporary power, 189
Temporary water supplies, 189
Tennessee Valley Authority, 92
Termination, 245
Three-part specification section format, 57
Time:
   cards, 180
   construction, 36, 192, 239, 243
   contract, **194**
   and cost monitoring, **185**
   disputes, 230
   of essence, **195, 274**
   extensions, 239
   and materials contract, 7
   monitoring, **185**
   performance, **154**
   preconstruction conference, 160
   prefinal inspection, **261**
   records, 180
   submittal of bids, **142**
Title VII, Equal Employment Opportunity, 106
Topography, impact on bidding, 214
Tornado, 32

Tracking costs and expenses, 179
Traffic control, 98
Transactions, on-site, **182**
Transfer of risk, 22
Transmittal of information, 170
Transmittal letters, 179
Treasury Department Circular 570, 143
Trench and excavation safety, **129,** 130
Types of drawings comprising construction
   contract, **86**

UBC, see Uniform Building Code (UBC)
UL, see Underwriters Laboratory (UL)
Unabsorbed home office overhead, 195, 245
Unabsorbed overhead, 232, 238
Unbalanced bid, 165
Unbalanced schedule of values, 165
Uncompleted Punch List items, **277**
Unconditional release, 205, 276
Unconditional waiver, 276
Uncorrectable error in bid, 19
Underabsorbed home office overhead, 233
Underground work, 26
Underwriters Laboratory (UL), 83
Unexcusable error, 15, 19
Unforeseen site conditions, 22, 35, 241, **250**
Unforeseen underground conditions, 22, **196,**
   198
Uniform Building Code (UBC), 79, 84
Uniform Plumbing Code, 79, 85
Uniform System for Building Specifications, 52
Union problems, 37
Unions, 110
U.S. Army Corps of Engineers, 52
U.S. Corps of Engineers, 102
U.S. Department of Labor, 108, 109
U.S. Government Superintendent of Documents,
   17
U.S. Navy (NAVFAC), 52
Unit price(s), 6, 7, 141
   vs. lump-sum bids, **140**
Unit price bidding, 7, 140, 149
Unit price contracts, **6,** 140, 141
   additions and deletions, 140–142
Unknown physical conditions, 197
Unknown subsurface conditions, 242
Unqualified personnel delay, 231
Unusually severe weather conditions, 242
Unusual site conditions, **250**
Use of schedules and graphics, 189

INDEX

Value engineering, **210**, **214**
   analysis phase, **218**
   change order, 213
   considerations, 211
   by contractor, **213**
   costs, 211
   definition, **210**, **211**
   development phase, **219**
   function, 211
   fundamentals, **211**
   incentive, 210, 211
   information phase, **217**
   measurement of value, 212
   methodology, **215**
   philosophy of value, **213**
   presentation of follow-up Phase, **220**
   speculation phase, 218
   types of recommendations, **213**
   value, 212
   worth, 212
Value Engineering Change Proposals (VECP), **213**, **221**
Value Engineering Proposal (VEP)(s), **213**, **215**, **219**, **220**
Value Engineering Job Plan, 215
VECP, **see** Value Engineering Change Proposal (VECP)
VEP, **see** Value Engineering Proposal (VEP)
Voluntary compensation, 282

Wage rates, 168
   Department of Labor, 14
   federal, 47
   prevailing, 12
   state, 47
Waiver(s):
   conditional and unconditional, 276
   of disputes, claims, and protests, 15
   of liability, **270**
   lien, 276
   of notice provision, 205
   of rights, 15
Warranties, **158**
Warranty, commencement, 265
Warranty dates, 261, 270
Water and sewer projects, sample documents, 303
Water systems, temporary, 190
WBE, *see* Woman-owned business enterprise (WBE)
Weather-caused delays, 145
Weather conditions, 180, 242
   normal, 32
   unusually severe, 242
WIC, *see* Woodwork Institute of California (WIC)
William–Steiger Occupational Safety and Health Act of 1970, 118
Wire Reinforcement Institute (WRI), 83
Withdrawal of bid, 19
Withheld funds from final payment, **269**
Withholding for uncompleted punch list items, 206
Woman-owned business enterprise (WBE), 14, 109
Woodwork Institute of California (WIC), 83
Work:
   conforming, 195
   beyond contract limits, 206
   by contractor, **261**
   directive change, **204**
   within or adjacent to navigable waterways, **99**
   nonconforming, 195, 207
   performed under protest, **226**
   under protest, 226
Worker productivity, 245
Workers' compensation insurance, **113**, 116, 281
   rates and safety, 113
Working days, 194
Work Zone Traffic Control, Standards and Guidelines, 128
WRI, *see* Wire Reinforcement Institute (WRI)